Mechanisms of inorganic reactions in solution

Consulting Editor

P. SYKES, M.Sc., Ph.D.,

Fellow of Christ's College
University of Cambridge

Other titles in the European Chemistry Series

BANWELL: Fundamentals of Molecular Spectroscopy

BARNARD: Theoretical Basis of Inorganic Chemistry

BARNARD & CHAYEN: Modern Methods of Chemical Analysis

BARNARD & MANSELL: Fundamentals of Physical Chemistry

BRENNAN & TIPPER: A Laboratory Manual of Experiments in Physical Chemistry

BU'LOCK: The Biosynthesis of Natural Products

CHAPMAN: Introduction to Lipids

HALLAS: Organic Stereochemistry

KEMP: Practical Organic Chemistry

KREBS: Fundamentals of Inorganic Crystal Chemistry

SUTTON: Electronic Spectra of Transition Metal Complexes

WILLIAMS & FLEMING: Spectroscopic Methods in Organic Chemistry

WILLIAMS & FLEMING: Spectroscopic Problems in Organic Chemistry

Mechanisms of inorganic reactions in solution

AN INTRODUCTION

D. BENSON, B.Sc., Ph.D., A.R.I.C.
Widnes Technical College

McGRAW-HILL · LONDON

New York · Sydney · Toronto · Mexico · Johannesburg · Panama

Published by

McGRAW-HILL Publishing Company Limited
MAIDENHEAD · BERKSHIRE · ENGLAND

94074

Printed and bound in Great Britain

To my wife and my children

Preface

My intention in writing this book is to provide, at an introductory level and within the compass of a small volume, an account of the mechanisms of inorganic reactions in homogeneous solution. In this respect I have kept in mind the needs of both the senior undergraduate and the post-graduate student beginning work in this field. For the benefit of the latter I have documented the material by literature references. The undergraduate student, having precious little time for reading the original papers, can, if he so wishes, extend the coverage of the book by supplementary reading chosen from the bibliographies provided at the end of each chapter. Since the inclusion of inorganic mechanism in the teaching curricula of universities and colleges is a fairly recent development, I hope also that the book may be of some use to professional chemists. A much more detailed account of the subject is given by F. Basolo and R. G. Pearson in *Mechanisms of Inorganic Reactions*, second edition, Wiley, 1967; unfortunately, the second edition appeared too late for me to include detailed references to it in my own book.

The plan of the book is as follows. The first chapter is meant to serve as a short refresher course on the general principles of reaction rate, mechanism, and related topics. However, an elementary knowledge of kinetics is assumed. The second chapter deals with substitution reactions of metal complexes—a vast field of current research as a glance at the book by Basolo and Pearson will show. Both six-coordinated and four-coordinated complexes are considered. Chapter 3 is concerned with oxidation–reduction reactions of metal ions. Chapters 4 and 5 cover reactions of oxoanions and free radical reactions, respectively, and the last chapter comprises a brief account of some protolytic reactions.

In preparing this book I would like to record my debt to a number of excellent reviews on various aspects of the subject, particularly those by Professors F. Basolo, J. Halpern, R. G. Pearson, and N. Sutin. I should also like to express my thanks to a number of authors for helpful discussions and for their kindness in allowing me to make use of diagrams from their publications: to them acknowledgement is made on the appropriate pages.

Finally, I would like to thank Mrs E. Geraghty, for her efficient typing of a tricky manuscript, and Dr L. H. Sutcliffe who, as well as reading the whole book in draft, offered advice and encouragement.

D. BENSON

Contents

CHAPTER

1. Introduction

This chapter is intended to serve as an introduction to some of the basic issues that will recur in the later sections of the book. Since the main approach to mechanism is the kinetic one, emphasis is placed on rate measurements and their interpretation. To supplement this necessarily brief coverage, the reader is referred to the works given in the selected bibliography on p. 17.

Theory of reaction rates

According to the theory of absolute reaction rates, the reactants in a chemical reaction are in equilibrium with an energetic species known as the *activated complex*. The thermal energy required to form such a complex is referred to as the *activation energy*. The activated complex occupies an energy level called the *transition state* which occurs at the point of highest energy on the path from one stable configuration to another (Fig. 1.1). Quantitatively the rate of passage from one stable state to the next can be derived from a statistical thermodynamic approach. If in the generalized bimolecular reaction

$$\mathrm{A} + \mathrm{B} \overset{K^{\ddagger}}{\rightleftharpoons} \mathrm{X}^{\ddagger} \rightarrow \mathrm{M} + \mathrm{N}$$

$X^{\ddagger}$ represents the activated complex whose equilibrium concentration is fixed by

$$[X^{\ddagger}] = K^{\ddagger}[A][B]$$

then the rate of formation of products is given by

$$\mathrm{d}[M]/\mathrm{d}t = (\boldsymbol{k}T/\boldsymbol{h})[X^{\ddagger}] = (\boldsymbol{k}T/\boldsymbol{h})\, K^{\ddagger}[A][B] = k_2[A][B] \qquad (1.1)$$

where k_2 is the second-order rate constant. The term $\boldsymbol{k}T/\boldsymbol{h}$, in which $\boldsymbol{k}$ and $\boldsymbol{h}$ are the Boltzmann and Planck constants, respectively, is a universal constant for all chemical reactions. An extension of this relationship is got by substituting the appropriate thermodynamic functions for $K^{\ddagger}$ when it follows that

$$d[M]/dt = (kT/h)\exp(-\Delta G^{\ddagger}/RT)[A][B] \quad (1.2)$$

or

$$d[M]/dt = (kT/h)\exp(\Delta S^{\ddagger}/R)\exp(-\Delta H^{\ddagger}/RT)[A][B] \quad (1.3)$$

Here $\Delta G^{\ddagger}$, $\Delta S^{\ddagger}$, and $\Delta H^{\ddagger}$ are the standard activation free energy, entropy, and enthalpy, respectively. For a reaction in solution, the

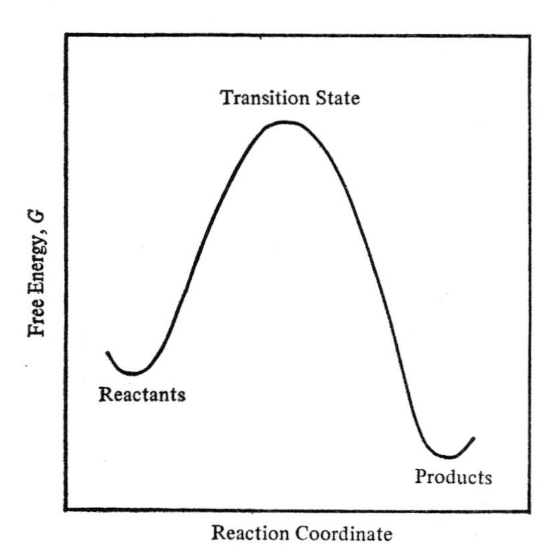

Fig. 1.1. Reaction profile showing position of transition state.

energy of activation is greater than $\Delta H^{\ddagger}(E = \Delta H^{\ddagger} + RT)$. However, since E is usually of the order of 10–20 kcal mole^{-1} whereas RT is 0·6 kcal mole^{-1} at 25°, the difference is commonly ignored.

The entropy of activation is a measure of the total entropy changes taking place in the reactants and the solvent on formation of the activated complex and, as such, its sign and magnitude are determined chiefly by the charge of the activated complex relative to the charges of the reactants. For reactions between oppositely-charged ions, the activated complex will have a lower charge than the reactants and, as a consequence, will be less solvated (i.e., less 'ordered'). Since the formation of the activated complex is accompanied by an increase in 'disorder', the $\Delta S^{\ddagger}$ value will be positive. For a reaction between ions of like charge $\Delta S^{\ddagger}$ will be negative. Quantitatively the electrostatic contribution to the entropy of activation is given approximately by $-10Z_AZ_B$ cal deg^{-1} mole^{-1}, for reaction between ions of charges Z_A and Z_B. The numerous exceptions to this simple electrostatic approach point to the existence of other effects less clearly defined. An interesting discussion of the positive entropies of activation found for some oxidation–reduction reactions of metal ions is given in a paper by Higginson.[1]

Strictly, for reactions of ions in solution, the foregoing treatment of reaction rate should be modified to take into account deviations from ideal behaviour. The equilibrium between reactants and activated complex is more correctly expressed in terms of activities such that

$$K_a^{\ddagger} = a_{X^{\ddagger}}/a_A a_B = ([X^{\ddagger}]/[A][B])(\gamma_{X^{\ddagger}}/\gamma_A \gamma_B) \tag{1.4}$$

where γ's represent activity coefficients. Since the *rate* of reaction is proportional to the *concentration* of activated complexes (eq. (1.1)), it follows that the experimental rate constant (k_2) depends upon the ratio of activity coefficients

$$k_2 = (\boldsymbol{kT/h})\, K_a^{\ddagger}(\gamma_A \gamma_B/\gamma_{X^{\ddagger}}) \tag{1.5}$$

At infinite dilution the γ's become unity and the limiting rate constant (k_0) is given by $k_0 = (\boldsymbol{kT/h}) K_a^{\ddagger}$. Thus

$$k_2 = k_0(\gamma_A \gamma_B/\gamma_{X^{\ddagger}}) \tag{1.6}$$

Due to ion-atmosphere effects, changes in solute concentration bring about alterations in activity coefficients and thus the rate constant varies with the total ionic strength. Quantitatively the effect of ionic strength (μ) on an activity coefficient is governed by the Debye–Hückel equation

$$-\log \gamma_i = \frac{Z_i^2 \beta \mu^{1/2}}{1 + \alpha\mu^{1/2}} \tag{1.7}$$

where Z_i is the charge of the ion, and α and β are constants (α is dependent on the closest distance of approach of the ith ion to another ion). Writing the activity coefficients in eq. (1.6) in terms of μ gives rise to the following relationship between the rate constant and ionic strength

$$\log \frac{k_2}{k_0} = \frac{2Z_A Z_B \beta \mu^{1/2}}{1 + \alpha\mu^{1/2}} \tag{1.8}$$

Consequently there is a linear relationship between log k_2 and $\mu^{1/2}/(1 + \alpha\mu^{1/2})$ and the corresponding plot has a slope of $2\beta Z_A Z_B$ ($\sim Z_A Z_B$ since $\beta = 0{\cdot}51$ at 25° in aqueous solution). Furthermore, extrapolation to $\mu = 0$ yields a value for k_0. The treatment is valid only for dilute solutions (less than 0·01 M for uni-univalent electrolytes) and in the absence of complex formation. In practice, as a precaution against unwanted ionic strength effects, the kinetics of a solution reaction are studied for a reaction medium of constant ionic strength. For example, oxidation–reduction reactions of metal ions are normally performed in perchloric acid solution (to reduce the risk of complexing between the metal ions and the medium) in the presence of a constant and large excess of sodium perchlorate. The use of a constant excess of an inert salt effectively swamps variations in ionic strength that would otherwise occur due to changes in reactant concentrations, acidity, etc. Unfortunately a general lack of appreciation of ionic strength effects invalidates much of the early literature on solution kinetics.

Rate laws and mechanism

The terms *intermediate* and activated (or transition state) complex should not be confused: an intermediate is understood as being a state of some stability occupying an energy minimum between reactants and products (Fig. 1.2). Reactions can be designated as simple or complex depending upon whether only one transition state intervenes between reactants and products or whether intermediates are formed and, as a result, more than one transition state occurs. In this context a reaction mechanism can be defined as a description of all the transition states and intermediates involved between reactants and products. Naturally the most decisive evidence for intermediate formation occurs in those circumstances where the intermediate can be isolated from the reaction mixture and then characterized. Although this can often be achieved for organic systems, isolation of the generally short-lived inorganic intermediate is rarely possible. Then the presence of the transient species can only be inferred by indirect means. Spectroscopic techniques, particularly ultraviolet and visible spectrophotometry, have been much used in this respect. Electron spin resonance has been made use of occasionally. Sometimes physical detection of a reactive intermediate is supported by evidence from kinetics; more often indication of the existence of an intermediate rests entirely on kinetic evidence.

For a simple reaction the composition of the activated complex follows directly from the form of rate law, and the thermodynamic properties may be calculated from the variation of the rate constant with temperature using the Arrhenius equation

$$d(\ln k)/dT = E/RT^2 \tag{1.9}$$

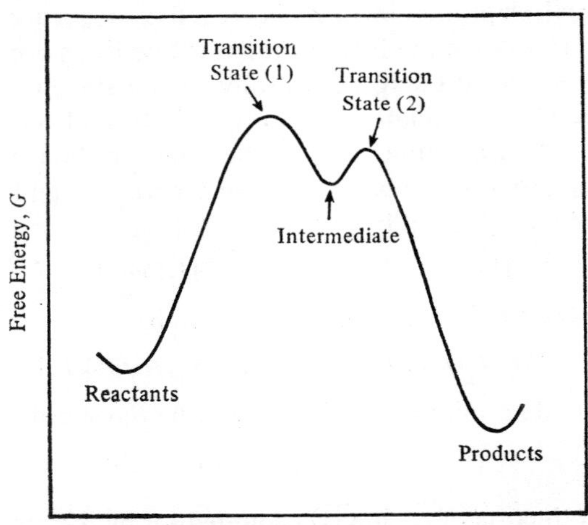

Fig. 1.2. Reaction profile showing position of intermediate and transition states.

together with eq. (1.3) on p. 2. To obtain information on the structure of the activated complex is much more difficult. Occasionally entropies of activation can provide such evidence. When the entropy of reaction (ΔS) approximates to $\Delta S^{\ddagger}$ it suggests that the transition state closely resembles the products. Conversely, if there is a serious disparity between the ΔS and $\Delta S^{\ddagger}$ values, as there is in the case of the reaction

$$[Co(NH_3)_5Cl]^{2+} + OH^- \rightarrow [Co(NH_3)_5OH]^{2+} + Cl^-$$

where $\Delta S = -1{\cdot}0$ and $\Delta S^{\ddagger} = +37$ cal deg^{-1} mole^{-1}, then the reaction must proceed by an indirect path. In this example the lack of correlation is convincing evidence against a direct bimolecular process. A point that must be stressed for reactions in solution concerns the uncertain role of the solvent in the transition state. In dilute solutions the concentration of the solvent cannot be varied and the order with respect to the solvent is always unknown. As a result the solvent content of the activated complex is impossible to evaluate.

A reaction whose rate law shows agreement with the stoichiometric equation can be either simple or complex in nature. On the other hand, incompatibility of the rate law with the stoichiometric equation is a clear indication that the reaction is complex. If, in a complex reaction, the individual steps making up the reaction sequence are not of comparable rates then the form of the rate law leads normally to the location of the slowest stage. Naturally such a stage determines the overall rate. Once evaluated, the composition and thermodynamic properties of the transition state can be found in the usual way. For example, the reaction between hydrogen peroxide and iodide ions has a stoichiometry given by

$$H_2O_2 + 3I^- + 2H^+ \rightarrow 2H_2O + I_3^- \quad (1)$$

but the rate law (at low acidities) is found to be[2]

$$-d[H_2O_2]/dt = d[I_3^-]/dt = k_1[H_2O_2][I^-]$$

Thus the rate-determining step is likely to be the simple process

$$H_2O_2 + I^- \rightarrow H_2O + IO^- \qquad \text{slow} \quad (2)$$

The final production of I_3^- ion is accounted for by the rapid reversible steps

$$IO^- + H^+ \rightleftharpoons HOI \quad (3)$$

$$HOI + H^+ + I^- \rightleftharpoons H_2O + I_2 \quad (4)$$

$$I^- + I_2 \rightleftharpoons I_3^- \quad (5)$$

Summation of eqs. (2) to (5) yields the stoichiometric equation, eq. (1).

Frequently the rate law assumes the form of a linear combination of several independent terms. In this case the individual terms represent contributions from different reaction paths. Each path has its own characteristic activated complex whose composition follows from the corresponding term in the rate law. An illustration of this is to be found in the kinetics of the peroxide–iodide reaction at high acidities, under which conditions a second term appears in the rate expression which now assumes the form

$$d[I_3^-]/dt = k_1[H_2O_2][I^-] + k_2[H_2O_2][I^-][H^+]$$

The occurrence of this hydrogen-ion dependent term suggests that there are two independent and competing steps of similar rate.

Two chief ambiguities arise, for solution reactions, in the interpretation of rate data. The first is associated with the inadequacy of the rate law in defining the composition of the activated complex whilst providing no indication of the species from which the transition state is derived. This is particularly relevant to the situation where one or a series of rapid equilibria precede the rate-controlling step. The acid-dependent path of the peroxide–iodide reaction provides an example of this.* There are a number of possibilities:

either

$$H_2O_2 + H^+ + I^- \rightarrow H_2O + HOI \qquad \text{slow} \tag{6}$$

or

$$H^+ + I^- \rightleftharpoons HI \qquad \text{rapid equilibrium}$$
$$H_2O_2 + HI \rightarrow H_2O + HOI \qquad \text{slow}$$

or

$$H^+ + H_2O_2 \rightleftharpoons H_3O_2^+ \qquad \text{rapid equilibrium}$$
$$H_3O_2^+ + I^- \rightarrow H_2O + HOI \qquad \text{slow}$$

Kinetically these three steps are indistinguishable. However, there is considerable prejudice against a reaction proceeding via a termolecular process like (6) since the existence of three-body collisions in solution is statistically improbable. Even if the relative concentrations of HI and $H_3O_2^+$ could be assessed, it would not be possible to decide which of the remaining processes were operative because a higher concentration of a species does not necessarily make a path involving that species kinetically significant.

A second ambiguity is noted when a reaction proceeds through two activated complexes formed consecutively. Here again the rate law serves to define the compositions of the transition states but is incapable of shedding any light as to the order in which they are formed. An interesting example of the mechanistic significance of this is given at the end of the next section.

Stationary-state approximation

For some complex reactions the isolation of one or two rate-controlling stages proves impossible and, because the rate is not governed

* A further, and classic, example occurs in the ammonium cyanate-urea conversion. The mechanism of this reaction is discussed in detail in the book by Frost and Pearson (see bibliography).

by any one step, the rate law is complicated. The important group of *chain reactions* fall into this category. In these circumstances the most that can be done is to postulate a mechanism and check then to see if it leads to an expression in agreement with the empirical rate law. Although non-agreement with the observed kinetics necessarily means rejection of a postulated mechanism, a mechanism cannot be considered correct solely on the grounds of compatibility with the rate law. Frequently more than one sequence of intermediates are acceptable kinetically; then the choice between them must rely on non-kinetic evidence (e.g., energetics) or on comparisons with analogous systems. Often the decision is largely intuitive.

An invaluable aid in comparing postulated mechanisms with empirical rate data is the *steady-state* (*or stationary-state*) *approximation* (Bodenstein, 1913). This method is applicable only to intermediates that are thermodynamically unstable and, consequently, are not present at appreciable concentrations. In these instances it is customary to assume that, after a short time (relative to the half-life of the reaction) has elapsed, a steady (or stationary) state is attained and, provided this is maintained, the concentrations of unstable intermediates are constant. Or, in other words, during the steady state the rates of change of the concentrations of the intermediates with time are so small as to be equated to zero. On this principle the concentrations of all active intermediates can be eliminated from the rate expression.

As an illustration of the steady-state principle in operation, consider the reaction between cobalt(III) and chromium(III) in perchloric acid solutions as catalysed by small amounts of silver(I) ion. The postulated mechanism is [3]

$$\mathrm{Co(III)} + \mathrm{Ag(I)} \underset{k_2}{\overset{k_1}{\rightleftharpoons}} \mathrm{Co(II)} + \mathrm{Ag(II)}$$

$$\mathrm{Cr(III)} + \mathrm{Ag(II)} \xrightarrow{k_3} \mathrm{Cr(IV)} + \mathrm{Ag(I)}$$

$$\mathrm{Cr(IV)} + \mathrm{Ag(II)} \xrightarrow{k_4} \mathrm{Cr(V)} + \mathrm{Ag(I)}$$

$$\mathrm{Cr(V)} + \mathrm{Ag(II)} \xrightarrow{k_5} \mathrm{Cr(VI)} + \mathrm{Ag(I)}$$

The reaction was followed spectrophotometrically by observing the rate of appearance of Cr(VI) from the increase of absorbance at 475 mμ, at which wavelength absorptions by Co(II), Co(III), and Cr(III) are slight. Silver(II), an intermediate in the reaction, also absorbs strongly at 475 mμ but, as it is present in very low concentrations, the observations are not affected. The rate of formation of Cr(VI) is given by

$$d[Cr(VI)]/dt = k_5[Cr(V)][Ag(II)] \quad (1.10)$$

By assuming the steady-state approximation, the rates of formation of the intermediates, Cr(V) and Cr(IV), can be set equal to zero. Thus

$$d[Cr(V)]/dt = k_4[Cr(IV)][Ag(II)] - k_5[Cr(V)][Ag(II)] = 0 \quad (1.11)$$

and

$$d[Cr(IV)]/dt = k_3[Cr(III)][Ag(II)] - k_4[Cr(IV)][Ag(II)] = 0 \quad (1.12)$$

This permits the concentrations of the intermediates to be expressed in terms of the reactants

$$[Cr(IV)] = k_3[Cr(III)]/k_4$$

and

$$[Cr(V)] = k_4[Cr(IV)]/k_5 = k_3[Cr(III)]/k_5$$

Equation (1.10) may then be transformed into

$$d[(Cr(VI)]/dt = k_3[Cr(III)][Ag(II)] \quad (1.13)$$

Also

$$\begin{aligned} d[Ag(II)]/dt = k_1[Co(III)][Ag(I)] \\ - k_2[Co(II)][Ag(II)] - k_3[Cr(III)][Ag(II)] \\ - k_4[Cr(IV)][Ag(II)] - k_5[Cr(V)][Ag(II)] = 0 \end{aligned}$$

From eqs. (1.11) and (1.12)

$$k_3[Cr(III)][Ag(II)] = k_4[Cr(IV)][Ag(II)] = k_5[Cr(V)][Ag(II)]$$

therefore

$$\begin{aligned} d[Ag(II)]/dt = k_1[Co(III)][Ag(I)] \\ - k_2[Co(II)][Ag(II)] - 3k_3[Cr(III)][Ag(II)] = 0 \end{aligned}$$

That is

$$[Ag(II)] = \frac{k_1[Co(III)][Ag(I)]}{k_2[Co(II)] + 3k_3[Cr(III)]}$$

Finally, elimination of $[Ag(II)]$ from (1.13) yields the rate law

$$\frac{d[Cr(VI)]}{dt} = \frac{k_1k_3[Cr(III)][Ag(I)][Co(III)]}{k_2[Co(II)] + 3k_3[Cr(III)]}$$

This derived rate law is in agreement with the observed kinetics

Oxidations by peroxodisulphate ($S_2O_8^{2-}$) provide an example of the value of the steady-state approximation in handling complicated chain mechanisms.[4] Uncatalysed oxidations by peroxodisulphate are commonly described by rate laws showing a first-order dependence on $S_2O_8^{2-}$ but a zero-order dependence on the reducing agent. Examples are the oxidations of thiosulphate, formic acid, 2-propanol, and, under certain conditions, hydrogen peroxide. The aqueous decomposition of peroxodisulphate also is first-order in $S_2O_8^{2-}$. A proposed mechanism must take account of this. Furthermore, it must explain why, under these circumstances, the decomposition of $S_2O_8^{2-}$ increases on the addition of the reducing agent. The following scheme has been suggested in which S represents the oxidizable substrate, R is a radical derived from S, and P is a product of the reaction:

$$S_2O_8^{2-} \xrightarrow{k_1} 2SO_4^-$$

$$SO_4^- + H_2O \xrightarrow{k_2} OH + HSO_4^-$$

$$OH + S \xrightarrow{k_3} OH^- + R$$

$$R + S_2O_8^{2-} \xrightarrow{k_4} P + SO_4^{2-} + SO_4^-$$

$$R + SO_4^- \xrightarrow{k_5} P + SO_4^{2-}$$

The rates of decomposition of $S_2O_8^{2-}$ and substrate are given by

$$-\mathrm{d}[S_2O_8^{2-}]/\mathrm{d}t = k_1[S_2O_8^{2-}] + k_4[R][S_2O_8^{2-}] \tag{1.14}$$

$$-\mathrm{d}[S]/\mathrm{d}t = k_3[OH][S] \tag{1.15}$$

The rates of formation of the radicals SO_4^-, R, and OH are given by

$$\mathrm{d}[SO_4^-]/\mathrm{d}t = 2k_1[S_2O_8^{2-}] - k_2[SO_4^-] + k_4[R][S_2O_8^{2-}] - k_5[R][SO_4^-] \tag{1.16}$$

$$\mathrm{d}[R]/\mathrm{d}t = k_3[OH][S] - k_4[R][S_2O_8^{2-}] - k_5[R][SO_4^-] \tag{1.17}$$

and

$$\mathrm{d}[OH]/\mathrm{d}t = k_2[SO_4^-] - k_3[OH][S] \tag{1.18}$$

Since, by the steady-state approximation, the concentrations of these radicals are constant

$$\mathrm{d}[SO_4^-]/\mathrm{d}t = \mathrm{d}[R]/\mathrm{d}t = \mathrm{d}[OH]/\mathrm{d}t = 0 \tag{1.19}$$

Therefore from eqs. (1.19) and (1.18) it follows that

$$k_2[SO_4^-] = k_3[OH][S]$$

or

$$[S] = k_2[SO_4^-]/k_3[OH]$$

which may be substituted into eq. (1.15) giving

$$-\mathrm{d}[S]/\mathrm{d}t = k_2[SO_4^-] \tag{1.20}$$

or into eq. (1.17) giving

$$\mathrm{d}[R]/\mathrm{d}t = k_2[SO_4^-] - k_4[R][S_2O_8^{2-}] - k_5[R][SO_4^-] = 0 \tag{1.21}$$

Addition and subtraction of eqs. (1.16) and (1.21) produce, respectively

$$k_1[S_2O_8^{2-}] = k_5[R][SO_4^-] \tag{1.22}$$

and

$$k_2[SO_4^-] = k_1[S_2O_8^{2-}] + k_4[R][S_2O_8^{2-}] \tag{1.23}$$

Inspection of eqs. (1.14), (1.20), and (1.23) reveals that

$$-\mathrm{d}[S_2O_8^{2-}]/\mathrm{d}t = -\mathrm{d}[S]/\mathrm{d}t$$

Also from eq. (1.22)

$$[SO_4^-] = k_1[S_2O_8^{2-}]/k_5[R]$$

Elimination of $[SO_4^-]$ in eq. (1.23) leads to

$$k_1 + k_4[R] = k_1k_2/k_5[R]$$

which on rearrangement becomes

$$k_4[R]^2 + k_1[R] - k_1k_2/k_5 = 0$$

$[R]$ is got by solving this quadratic

$$[R] = [-k_1 \pm (k_1^2 + 4k_1k_2k_4/k_5)^{1/2}]/2k_4$$

Finally substitution for $[R]$ in eq. (1.14) gives

$$\begin{aligned}\text{rate} &= -\mathrm{d}[S_2O_8^{2-}]/\mathrm{d}t = -\mathrm{d}[S]/\mathrm{d}t \\ &= \tfrac{1}{2}[k_1 \pm (k_1^2 + 4k_1k_2k_4/k_5)^{1/2}][S_2O_8^{2-}]\end{aligned}$$

If k_1 is assumed to be small, this rather formidable expression simplifies to

$$\text{rate} = -\mathrm{d}[S_2O_8^{2-}]/\mathrm{d}t = -\mathrm{d}[S]/\mathrm{d}t = (k_1k_2k_4/k_5)^{1/2}[S_2O_8^{2-}]$$

As observed, the rate is independent of the concentration of substrate but is first-order in $S_2O_8^{2-}$ concentration. Furthermore, rearrangement of eq. (1.14)

$$-d[S_2O_8^{2-}]/dt = (k_1 + k_4[R])[S_2O_8^{2-}]$$

shows that the different rates observed for various substrates are due to variations in the concentrations of radicals derived from the substrates.

A revealing paper by Haim[5] is devoted to a discussion of the mechanistic complexities of the chromium(II) + vanadium(III) reaction. The experimental results, for perchloric acid media, are summarized by the expression

$$\text{rate} = \frac{q[V^{3+}][Cr^{2+}]}{r + [H^+]}$$

The rate law has two limiting forms: at low hydrogen-ion concentrations

$$\text{rate} = q[V^{3+}][Cr^{2+}]/r$$

and at high $[H^+]$

$$\text{rate} = q[V^{3+}][Cr^{2+}]/[H^+]$$

Thus the reaction must proceed via two activated complexes which can be identified as $[VCr^{5+}]^{\ddagger}$ and $[V(OH)Cr^{4+}]^{\ddagger}$. The mechanism originally proposed[6] had as an intermediate the binuclear species $V(OH)Cr^{4+}$ in which V^{3+} and Cr^{2+} are linked by a hydroxo bridge

$$V^{3+} + Cr^{2+} + H_2O \underset{k_2}{\overset{k_1}{\rightleftharpoons}} V(OH)Cr^{4+} + H^+$$

$$V(OH)Cr^{4+} \xrightarrow{k_3} V^{2+} + CrOH^{2+}$$

$$CrOH^{2+} + H^+ \rightleftharpoons Cr^{3+} + H_2O \qquad \text{rapid equilibrium}$$

Assuming steady-state conditions to apply to $V(OH)Cr^{4+}$

$$\text{rate} = \frac{-d[V^{3+}]}{dt} = \frac{-d[Cr^{2+}]}{dt} = \frac{k_1k_3[V^{3+}][Cr^{2+}]}{k_3 + k_2[H^+]}$$

and the scheme agrees with experiment. Furthermore, by making use of the observed parameters, q and r, k_1 and k_3/k_2 can be calculated as 5·76 M^{-1} s^{-1} and 0·108 M, respectively, at 25° ($q = k_1k_3/k_2$ and

$r = k_3/k_2$). However, the kinetic law is equally in accord with the mechanism

$$V^{3+} + H_2O \rightleftharpoons VOH^{2+} + H^+ \qquad \text{rapid equilibrium, } K$$

$$VOH^{2+} + Cr^{2+} \underset{k_5}{\overset{k_4}{\rightleftharpoons}} V(OH)Cr^{4+}$$

$$V(OH)Cr^{4+} + H^+ \xrightarrow{k_6} V^{2+} + Cr^{3+} + H_2O$$

In this alternative mechanism the same activated complexes are involved but $[CrV^{5+}]^{\ddagger}$ is derived from $V(OH)Cr^{4+} + H^+$ ($[CrV(H_2O)^{5+}]^{\ddagger}$ is equivalent to $[CrV^{5+}]^{\ddagger}$) and $[V(OH)Cr^{4+}]^{\ddagger}$ from $VOH^{2+} + Cr^{2+}$. The derived rate law

$$\text{rate} = \frac{k_4 k_6 K[V^{3+}][Cr^{2+}]}{k_5 + k_6[H^+]}$$

and the empirical rate law are again in agreement. From a knowledge of K, q, and r, k_4 and k_5/k_6 are calculated as $3{\cdot}12 \times 10^2$ M^{-1} s^{-1} and 0·108 M, respectively, at 25° ($q = k_4 K$ and $r = k_5/k_6$). In the first mechanism the acid-independent step involving $[VCr^{5+}]^{\ddagger}$ comes before the acid-dependent step involving $[V(OH)Cr^{4+}]^{\ddagger}$ whereas the opposite is true in the second mechanism.

The solvent and solvation of ions

The use of water as a medium for inorganic reactions is widespread. As a solvent it has obvious virtues but it is unfortunate that certain of its properties can prove troublesome in kinetic studies. One particular property, which has persistently complicated the interpretation of ligand substitution reactions, is the ability of water to function as a highly nucleophilic reagent such that the substitution of one ligand by another rarely takes place without the prior formation of an aquo complex. For example, in general the replacement of a ligand X^- in a cobalt(III) complex by a ligand Y^-, rather than taking place directly

$$[CoA_5X]^{2+} + Y^- \rightarrow [CoA_5Y]^{2+} + X^-$$

occurs through an aquation step

$$[CoA_5X]^{2+} + H_2O \xrightarrow{\text{slow}} [CoA_5(H_2O)]^{3+} + X^-$$

$$[CoA_5(H_2O)]^{3+} + Y^- \xrightarrow{\text{fast}} [CoA_5Y]^{2+} + H_2O$$

and as a consequence the overall rate of substitution becomes independent of the concentration of the entering reagent. Furthermore, an evaluation of the aquation step is itself difficult as a result of the inability to discriminate kinetically between a (S_N1) dissociative mechanism

$$[CoA_5X]^{2+} \underset{}{\overset{\text{slow}}{\rightleftharpoons}} [CoA_5]^{3+} + X^-$$

$$[CoA_5]^{3+} + H_2O \xrightarrow{\text{fast}} [CoA_5(H_2O)]^{3+}$$

and a (S_N2) solvolysis mechanism

$$[CoA_5X]^{2+} + H_2O \rightarrow [CoA_5(H_2O)]^{3+} + X^-$$

In addition, water can enter into acid–base reactions of the type

$$[CoA_5(H_2O)]^{3+} + H_2O \rightleftharpoons [CoA_5OH]^{2+} + H_3O^+$$

A further complication ensues from the lack of detailed information concerning the nature of metal ions in aqueous solutions. Although often written as such, it must be understood that metal cations are not bare ions but aquo complexes. Thus, for example, a reaction such as

$$Co^{3+} + Cl^- \rightarrow CoCl^{2+}$$

is only a convenient abbreviation for

$$[Co(H_2O)_n]^{3+} + Cl^- \rightarrow [Co(H_2O)_{n-1}Cl]^{2+} + H_2O$$

That is, the formation of the monochloro complex of cobalt(III) is not a direct electrostatic combination of two oppositely-charged ions but a substitution process in which the entering reagent displaces a coordinated water molecule from the inner coordination sphere of the metal ion. The actual number of water molecules composing the inner coordination sphere is known with certainty in a few instances only. In the absence of direct information it is normally assumed that this is the same as the usual coordination number of the metal ion (i.e., $n = 6$ for Co(III)). If this is the case, the replacement of a water molecule by an attacking nucleophile is likely to be an indirect process requiring, as a first stage, the expulsion of a water molecule from the hydration sphere. Conversely, if the normal coordination number is not satisfied then a vacant bonding orbital could act as a centre of attack for the incoming nucleophile and the process would then be a direct one.

Direct evidence on hydration is not easy to acquire. However, oxygen-18 experiments have clearly established the existence of the species $[Cr(H_2O)_6]^{3+}$.[7] On mixing solutions of Cr^{3+} and $H_2{}^{18}O$, a slow isotopic exchange takes place

$$[Cr(H_2O)_n]^{3+} + nH_2{}^{18}O \rightleftharpoons [Cr(H_2{}^{18}O)_n]^{3+} + nH_2O$$

which can be followed by sampling the oxygen-18 content of the solvent at various times by vacuum distillation. The isotopic exchange has a half-life of about 40 hours at 25°. For a given solution of Cr^{3+}, the number of 'unbound' water molecules can be calculated from the isotopic dilution which results within a short time of mixing with ^{18}O-containing solvent; the number of 'bound' water molecules is the difference between this value and the total number present. Ions of Cr^{3+} to molecules of bound water were found to be in the proportion of 1 to 6. The number of water molecules comprising the primary coordination shell of a metal ion has rarely been so exactly defined as for the case of Cr(III). The determination becomes more difficult as the exchange rate increases. However, by employing a flow technique, a value of 6 for the hydration number of Al^{3+} was obtained.[8] At the present time there is evidence for the existence of primary solvation in most metallic cations although the hydration numbers are still in some doubt. For a readable account of metal ions in aqueous solution the reader is referred to the book by Hunt.[9]

In an attempt to avoid complications (of the type discussed above) arising from the use of water as a solvent, there is an increasing interest in non-aqueous solvents as media for inorganic reactions. On practical grounds, the choice of a suitable non-aqueous solvent is considerably restricted by the rather poor solubility of most inorganic substances. The tendency is to make use of polar solvents like methanol, ethanol, dimethylformamide, and dimethylsulphoxide, although a number of studies have been made in non-polar solvents like acetic acid. However, the difficulties attending the use of aqueous systems are to a similar or greater extent present also in non-aqueous systems. In general, ionic association occurs to an increasing degree as the dielectric constant of the medium decreases and the uncertainties about the nature of the ion aggregates present can lead to difficulties in the interpretation of kinetic data. For polar solvents there is evidence to suggest direct solvolysis of complexes. Also there is little systematic knowledge about the nature of metal ions in non-aqueous media. In this context, it is probably true to say that kinetic studies of the reactions of metal ions in such media are practically worthless unless they are prefixed by a non-kinetic study of the species present.

Inert and labile complexes

The method most commonly adopted for the preparation of metal complexes is the substitution reaction. Thus $[Cu(NH_3)_4]SO_4$ and $K_3[Rh(C_2O_4)_3]$ can be synthesized by the following reactions:

$$[Cu(H_2O)_4]^{2+} + 4NH_3 \rightarrow [Cu(NH_3)_4]^{2+} + 4H_2O$$

$$[RhCl_6]^{3-} + 3C_2O_4{}^{2-} \rightarrow [Rh(C_2O_4)_3]^{3-} + 6Cl^-$$

Considerable differences in ease of ligand replacement can occur: the first reaction is virtually instantaneous at room temperature whereas to form the oxalato complex of Rh(III) requires boiling concentrated solutions of the reactants for several hours. Similarly the addition of chloride to Cu(II) solution results in the formation of a series of complexes by successive rapidly-established equilibria: an alteration in chloride concentration instantly adjusts the concentration of each chloro complex. In contrast, the saturation by hydrogen chloride of a hot aqueous solution of Cr(III) achieves nothing more than the separation of the compound $[Cr(H_2O)_6]Cl_3$ on cooling. Chloro complexes form only after a long period of time and eventually a mixed product, containing species of the type $[Cr(H_2O)_4Cl_2]Cl$, can be obtained by prolonged evaporation.

In the first systematic review of substitution reactions, Taube[10] proposed that, depending upon their reactivity, complexes should be described as *labile* or *inert*, and these terms are now in common usage. Those complexes which undergo ligand replacement within 1 minute (or so) at 25° and 0·1 M reactant concentration are arbitrarily termed labile; other less reactive complexes are referred to as inert. In the examples cited above it is apparent that the four-coordinated complexes of Cu(II) are labile, whereas the six-coordinated complexes of Rh(III) and Cr(III) are inert.

Major differences of behaviour occur also in the exchange reactions of the complex cyanides of Ni(II), Mn(III), and Cr(III)

	half-life
$[Ni(CN)_4]^{2-} + 4^{14}CN^- \rightleftharpoons [Ni(^{14}CN)_4]^{2-} + 4CN^-$	~30 sec
$[Mn(CN)_6]^{3-} + 6^{14}CN^- \rightleftharpoons [Mn(^{14}CN)_6]^{3-} + 6CN^-$	~1 hour
$[Cr(CN)_6]^{3-} + 6^{14}CN^- \rightleftharpoons [Cr(^{14}CN)_6]^{3-} + 6CN^-$	~24 days

although all three complexes are extremely stable in a thermodynamic sense (with high overall stability constants in the range 10^{22} to 10^{64}). Conversely, cobalt(III) ammine complexes like $[Co(NH_3)_6]^{3+}$ are energetically unstable in acid solution but resist decomposition in such conditions for several days at room temperature. Such complexes are unreactive or inert, but unstable. Evidently it is most important to distinguish carefully between instability and lability. From simple rate theory, the lability of a complex is dependent upon the activation energy, i.e., the difference in energy between the reactants and the activated complex; a low activation energy results in a fast reaction. Instability is decided by the difference between the free energies of the reactants and the products (loosely by the heat of reaction).

References

1. W. C. E. Higginson, D. R. Rosseinsky, J. B. Stead, and A. G. Sykes, *Disc. Faraday Soc.*, 1960, **29**, 49.
2. See, for example, H. A. Liebhafsky and A. Mohammad, *J. Amer. Chem. Soc.*, 1933, **55**, 3977.
3. J. B. Kirwin, P. J. Proll, and L. H. Sutcliffe, *Trans. Faraday Soc.*, 1964, **60**, 119.
4. D. A. House, *Chem. Rev.*, 1962, **62**, 185.
5. A. Haim, *Inorg. Chem.*, 1966, **5**, 2081.
6. J. H. Espenson, *Inorg. Chem.*, 1965, **4**, 1025.
7. J. P. Hunt and H. Taube, *J. Chem. Phys.*, 1951, **19**, 602.
8. H. H. Baldwin and H. Taube, *J. Chem. Phys.*, 1960, **33**, 206.
9. J. P. Hunt, *Metal Ions in Aqueous Solution*, Benjamin, 1963.
10. H. Taube, *Chem. Rev.*, 1952, **50**, 69.

Bibliography

General

S. W. Benson, *The Foundations of Chemical Kinetics*, McGraw-Hill, 1960.
E. S. Amis, *Kinetics of Chemical Change in Solution*, Macmillan, 1949.
E. L. King, *How Chemical Reactions Occur*, Benjamin, 1964.
A. A. Frost and R. G. Pearson, *Kinetics and Mechanism*, Second Edition, Wiley, 1961.
H. Eyring and E. M. Eyring, *Modern Chemical Kinetics*, Reinhold, 1963.

Experimental methods

D. R. Stranks, The Reaction Rates of Transitional Metal Complexes, in *Modern Coordination Chemistry* (ed. J. Lewis and R. G. Wilkins), p. 78, Interscience, 1960.

E. M. Eyring, Fast Reactions in Solution, in *Survey of Progress in Chemistry* (ed. A. F. Scott), Vol. 2, p. 57, Academic Press, 1964.
E. F. Caldin, *Fast Reactions in Solution*, Blackwell, 1964.

Volume 8 of *Technique of Organic Chemistry* (ed. S. L. Friess, E. S. Lewis, and A. Weissberger), Second Edition, Interscience, 1963 is entitled *Investigation of Rates and Mechanism of Reactions.* Part I gives a general account of Kinetic Methods. Of particular interest are the articles by R. Livingston on General Theory of Rate Processes (p. 15), Fundamental Operations and Measurements in Obtaining Rate Data (p. 55), and Evaluation and Interpretation of Rate Data (p. 109), and the account by J. F. Bunnett on The Interpretation of Rate Data (p. 177). Part II of the same work includes accounts of Rapid Reactions and Very Rapid Reactions. The latter topic is introduced by M. Eigen on p. 793.

2. Substitution reactions of metal complexes

Following organic terminology, inorganic chemists have found it convenient to divide substitution reactions of other elements than carbon into *nucleophilic* (S_N) and *electrophilic* (S_E):

$$[MX_n] + Y \rightarrow [MX_{n-1}Y] + X \qquad S_N$$

$$[MX_n] + M' \rightarrow [M'X_n] + M \qquad S_E$$

For this purpose, reactions which involve a change in oxidation number are excluded. Electrophilic substitution mechanisms are comparatively rare and will not be considered further (an example is the reaction between $[Co(NH_3)_5Cl]^{2+}$ and Hg^{2+}). The nucleophilic substitution reaction is a special type of general acid-base process, in which the metal ion functions as a Lewis acid (electron-pair acceptor) and the replacing ligand as a Lewis base (electron-pair donor).

S_N mechanisms are further subdivided into S_N1 dissociation (substitution, nucleophilic, unimolecular)

$$[MX_n] \underset{}{\overset{\text{slow}}{\rightleftharpoons}} [MX_{n-1}] + X$$

$$[MX_{n-1}] + Y \xrightarrow{\text{fast}} [MX_{n-1}Y]$$

$$\text{with rate} \propto [MX_n]$$

and S_N2 displacement (substitution, nucleophilic, bimolecular)

$$[MX_n] + Y \overset{\text{slow}}{\rightleftharpoons} [MX_nY] \xrightarrow{\text{fast}} [MX_{n-1}Y] + X$$

$$\text{with rate} \propto [MX_n][Y]$$

Most of the kinetic work on substitution reactions in recent years has been done in an endeavour to attach S_N1 and S_N2 labels. It should be stressed, however, that mechanisms other than these extremes are possible, and also that unequivocal direct evidence is difficult, if not impossible, to obtain.

An alternative classification which takes away undue emphasis on the molecularity of reaction has been proposed by Langford and Gray.[1] On their definition, ligand replacement reactions can proceed via (*a*) a *dissociative* path in which the leaving group is lost in the first (rate-determining) step, thus producing an intermediate of reduced coordination number which may be detectable on account of its varying reactivity to different reagents; or (*b*) an *associative* path in which the entering group adds on in the first (rate-determining) step, thus producing an intermediate of increased coordination number; a departure from second-order kinetics is to be expected in certain cases; or (*c*) a *concerted* path (or *interchange*) in which the leaving ligand is moving from the inner to the outer coordination sphere as the entering ligand is moving from the outer to the inner. A characteristic of interchange is thus the absence of an intermediate of modified coordination number. Dissociative and associative mechanisms are, of course, equivalent to S_N1 and S_N2, respectively; the term *association* is preferable to *displacement* which is descriptive of the overall process but not the primary step.

Using simple electrostatic arguments it is possible to speculate what will be the influence of changes in sizes and charges of the central metal ion, the entering group, the displaced group and the passive, non-replaced ligands on the rate of the three possible processes. However, such broad generalizations, founded as they are on a naïve model, must be treated with caution. Table 2.1 has been constructed on the basis that the important features of dissociative and associative mechanisms are bond-breaking and bond formation, respectively, whereas in an interchange process bond-breaking and bond formation have similar status. Assuming other factors remain constant, increasing the charge on the central metal atom will increase the strength of the metal–ligand bond and hinder a dissociative mechanism of charge separation, but will favour an associative process by aiding new bond formation with an entering group. Increased size of the central metal atom is expected to weaken the metal–ligand bond and also facilitate the accommodation of an entering group, hence an increase of rate is predicted irrespective of mechanism. As the rate-determining step in a dissociative process does not involve the entering group, it is obvious that changes in charge *or* size of the entering nucleophile will have no effect on the rate. In an associative process, increased charge of the substituent will increase the rate by promoting new bond formation, but increased size will impede bond formation.

Regardless of mechanism, it is to be expected that increase in (negative) charge of the leaving group will reduce the rate of substitution: bond-breaking is rendered more difficult, as is also bond

Table 2.1

Effect of sizes and charges on rates of dissociative and associative reactions

Changes made	Dissociative rate	Interchange rate	Associative rate
Increase positive charge of central atom	decrease	opposing effects	increase
Increase size of central atom	increase	increase	increase
Increase negative charge of entering group	no effect	increase	increase
Increase size of entering group	no effect	decrease	decrease
Increase negative charge of leaving group	decrease	decrease	decrease
Increase size of leaving group	increase	opposing effects	decrease
Increase negative charge of non-labile ligands	increase	opposing effects	decrease
Increase size of non-labile ligands	increase	opposing effects	decrease

formation due to the smaller effective positive charge on the central metal. An increase in size of the leaving group will make the rupture of the metal–ligand bond easier and assist a dissociative process; but the rate of reaction via an associative mechanism will be reduced due to difficulty in expanding the coordination number. Finally, the nature of the non-labile, unreplaced ligands can affect the rate. A greater negative charge on the non-labile ligands will help a dissociative process by repelling the reactive group but, since bond formation is made more difficult, will decrease the rate of an associative process. Steric strain resulting from increasing the size of the non-reactive ligands will assist a dissociative process but, by hindering the expansion of the coordination shell, will decrease the rate via an associative mechanism.

Indirect information on the mechanism of substitution has been obtained by investigating the effects of some of the changes summarized in Table 2.1. The following general conclusions can be drawn:

(*a*) Varying the size of the central metal atom or the charge of the leaving group is unable to assist in assignment of dissociative or associative mechanism since the response of the rate is the same on either process.
(*b*) Varying the charge and size of the entering group will, in principle, distinguish a dissociative mechanism.
(*c*) For the other possible variations given in Table 2.1, 'pure' dissociation or 'pure' association cannot be distinguished due to the possibility of reaction via a concerted interchange path, on which mechanism a net positive or negative rate response can be expected. If the assumption is made that the opposing effects noted for the interchange path cancel to a large extent, then tentative assignment of dissociative or associative mechanism can be made.

A. Octahedral complexes

Crystal-field theory has been applied with moderate success to the problem of relative reactivity in octahedral complexes.[2] Calculations, based upon the concept of crystal-field stabilization energy (CFSE), have revealed a pattern of kinetic behaviour in terms of the electronic configuration of the central metal ion. If the CFSE of the activated complex can be shown to be greater than that of the coordination compound, then on reaction via the transition state there will be a gain of CFSE which will result in a low activation energy and a rapid reaction. If the activated complex is less CF stabilized than the original substrate, there will be an increased activation energy and a lower rate of reaction. By taking the five-coordinated square pyramid and the seven-coordinated pentagonal bipyramid structures as approximations to the transition states adopted by an octahedral complex, reacting via a dissociative and associative mechanism respectively, the results collected in Table 2.2 are obtained. It is seen that systems which suffer a loss in CFSE, on *either* mechanism, are

d^3, d^8, and strong-field (low-spin) d^4, d^5, and d^6. Such configurations are predicted to give rise to inertness, while other configurations impart lability. This line of demarcation has been verified experimentally.

Although it is perhaps too much to expect from a simplified approach (based upon a number of rather drastic assumptions,

Table 2.2

Changes in CFSE (units Δ) upon changing a six-coordinated species to a five- or seven-coordinated species*

No. of d electrons	Weak field		Strong field	
	CN 5	CN 7	CN 5	CN 7
0	0	0	0	0
1	+0·06	+0·13	+0·06	+0·13
2	+0·11	+0·26	+0·11	+0·26
3	−0·20	−0·43	−0·20	−0·43
4	+0·31	−0·11	−0·14	−0·30
5	0	0	−0·09	−0·17
6	+0·06	+0·13	−0·40	−0·85
7	+0·11	+0·26	+0·11	−0·53
8	−0·20	−0·43	−0·20	−0·43
9	+0·31	−0·11	+0·31	−0·11
10	0	0	0	0

* Assuming retention of spin-multiplicity during reaction.

including the neglect of the charge of the central metal atom), the order of reactivity follows from the magnitude of the CFSE losses and is: $d^5 > d^4 > d^8 \sim d^3 > d^6$, e.g., Fe(III), Mn(II) > Cr(II), Mn(III) > Ni(II) ~ Cr(III) > Co(III), Fe(II). Certainly assignment of comparable reactivity to the d^3 and d^8 systems is incorrect. Complexes of Ni(II), in general, are considerably more labile than analogous Cr(III) complexes (compare, for example, the exchange rates of their complex cyanides, p.16).

A study of the dissociation rates of phenanthroline and terpyridyl complexes of Fe(II), Co(II), and Ni(II)[3] affords an example of correlations between d configuration and rate (see Table 2.3). For mono-phenanthroline and bis-terpyridyl complexes the rates of reaction decrease in the order high-spin $d^5 > d^7 > d^8 >$ low-spin d^6. The order of reactivity for mono-terpyridyl complexes is high-spin d^5 > high-spin d^6 > d^7 > d^8. The difference in behaviour of high- and low-spin d^6 complexes of Fe(II) is striking, and the greater lability of d^8 Ni(II) over low-spin d^6 Fe(II) is in accordance with the CFSE approach.

Table 2.3

Kinetic data for the dissociation of terpyridine (terpy) and phenanthroline (phen) complexes of bivalent metal ions (pH ~7 and 25°)*

Complex	E, kcal mole^{-1}	10^3k, min^{-1}	Electronic configuration	Change in CFSE (units Δ)†
(A) $[Mn\ phen]^{2+}$	10·4	very fast	d^5 (high-spin)	0
$[Co\ phen]^{2+}$	19·4	960	d^7	+0·11
$[Ni\ phen]^{2+}$	24·5	0·6	d^8	−0·20
$[Fe\ phen]^{2+}$	32·1	very slow	d^6 (low-spin)	−0·40
(B) $[Co\ terpy_2]^{2+}$	14·8	38	d^7	+0·11
$[Ni\ terpy_2]^{2+}$	20·8	0·1	d^8	−0·20
$[Fe\ terpy_2]^{2+}$	28·7	0·01	d^6 (low-spin)	−0·40
(C) $[Mn\ terpy]^{2+}$	12·3	very fast	d^5 (high-spin)	0
$[Fe\ terpy]^{2+}$	18·0	378	d^6 (high-spin)	+0·06
$[Co\ terpy]^{2+}$	20·2	6·0	d^7	+0·11
$[Ni\ terpy]^{2+}$	24·2	0·0015	d^8	−0·20

* From ref. (3).

† Assuming octahedral → square-pyramid structure. It may be more realistic to take values of CFSE changes shown as positive as equal to zero (ref. (2)).

It is necessary to emphasize that the electronic structure of the central metal ion is not the sole factor influencing the lability of metal complexes. Other factors of obvious importance are the charge and size of the central metal ion, which determine the strength of the metal–ligand bond. An illustration of charge affecting lability occurs in the isoelectronic series $[AlF_6]^{3-}$ > $[SiF_6]^{2-}$ > $[PF_6]^-$ ≫ SF_6, where increasing charge of the metal ion gives rise to an increase in metal–ligand bond strength and greater inertness. Thus slow reactions are observed among complexes, such as $[PF_6]^-$, $[AsF_6]^-$, $[SbCl_6]^-$, SF_6, SeF_6, and TeF_6, where the metal ion carries a high formal charge.

Similarly the elegant studies of Eigen on fast reactions show that in water exchanges of the type

$$[M(H_2O)_6]^{n+} + 6H_2\overset{*}{O} \rightleftharpoons [M(H_2\overset{*}{O})_6]^{n+} + 6H_2O$$

the rate of exchange decreases with increasing cationic charge in the order

$$[Na(H_2O)_n]^{+} > [Mg(H_2O)_n]^{2+} > [Al(H_2O)_6]^{3+}$$

If the size of the metal ion decreases but the charge remains constant then a diminution in reactivity takes place, e.g., in the exchange of hydrated metal ions: $Ba^{2+} > Sr^{2+} > Ca^{2+} \gg Mg^{2+} \gg Be^{2+}$, and $Cs^{+} > Rb^{+} > K^{+} > Na^{+} > Li^{+}$. The observation, that lability is favoured by a low charge-to-radius ratio, is shown to be consistent with substitution proceeding via a dissociative mechanism (cf. Table 2.1).

In the detailed discussion that follows we shall be concerned mainly with the substitution reactions of the low-spin complexes of Co(III) which has a d^6 electronic configuration. Such complexes are easily prepared and are sufficiently inert to be studied by conventional kinetic techniques. Some attention has been given, however, to the lesser-studied reactions of low-spin complexes of other d^6 systems, Rh(III), Ir(III), Pt(IV), and Fe(II), and also to the d^3 complexes of Cr(III) and high-spin complexes of d^8 Ni(II). The final section deals with the substitution of anionic ligands in the coordination shell of hydrated metal ions, a field of research requiring the use of fast reaction techniques.

Acid hydrolysis

Of the replacement reactions of octahedral coordination compounds the most widely studied are those in which aquo complexes are formed by *acid hydrolysis*

$$[MA_5X]^{n+} + H_2O \rightleftharpoons [MA_5H_2O]^{(n+1)+} + X^{-}$$

An alternative term to acid hydrolysis is *aquation*. It should be noted that acid hydrolysis implies no more than hydrolysis under acid conditions; the rate of the reaction may or may not be proportional to the concentration of acid. The greater bulk of work performed has been concerned with the acid hydrolysis of Co(III) complexes since it is known that most substitution processes of Co(III) in aqueous solution proceed indirectly via reaction with the solvent. ***Base hydrolysis***, to produce hydroxo complexes, is discussed on p. 40

$$[MA_5X]^{n+} + OH^- \rightarrow [MA_5OH]^{n+} + X^-$$

For those acid hydrolyses (at pH < 4) which show no acid dependence, the observed rate law is one of simple first-order

$$\text{rate} = k[MA_5X^{n+}]$$

This form of rate law is compatible with mechanism (*a*) or (*b*):

(*a*)
$$[MA_5X]^{n+} \underset{\text{fast}}{\overset{k_a,\ \text{slow}}{\rightleftharpoons}} [MA_5]^{(n+1)+} + X^-$$

$$[MA_5]^{(n+1)+} + H_2O \xrightarrow{\text{fast}} [MA_5H_2O]^{(n+1)+}$$

$$\text{rate} = k_a[MA_5X^{n+}]$$

(*b*)
$$[MA_5X]^{n+} + H_2O \overset{k_b,\ \text{slow}}{\rightleftharpoons}$$

$$[MA_5X.H_2O]^{n+} \xrightarrow{\text{fast}} [MA_5H_2O]^{(n+1)+} + X^-$$

$$\text{rate} = k_b[MA_5X^{n+}][H_2O] = k_b'[MA_5X^{n+}]$$

Recourse to other experimental data has to be made to distinguish between dissociation (*a*) and association (*b*). A particularly fruitful line of approach has been to study the effects resulting from the variation of non-replaceable groups present in the complex.

The most easily interpretable studies should be those designed to alter steric effects. It would be anticipated that 'crowding' of a reaction centre might favour a dissociation process (cf. Table 2.1) by effectively alleviating some degree of steric strain, but reduce the chance of association by discouraging attack of an approaching group. An example of the study of steric effects is the acid hydrolysis of Co(III) complexes of the type *trans*-$[Co(AA)_2Cl_2]^+$ (where AA represents a bidentate ligand derived from ethylenediamine). Some data are recorded in Table 2.4. Inspection of Table 2.4 reveals a correlation between rate and increased C-methyl substitution on ethylenediamine.[4] Furthermore, enlargement of the chelate ring from five to six, by the use of triethylenediamine instead of ethylenediamine, is seen to increase the rate drastically. This acceleration of rate is in accordance with a dissociative mechanism, in which the increasing steric strain is conducive to the formation of an intermediate of reduced coordination number.

Table 2.4

Rate constants for the acid hydrolysis of *trans*-$[Co(AA)_2Cl_2]^+$ at pH 1 and 25°
(replacement of first Cl^- only)*

$$[Co(AA)_2Cl_2]^+ + H_2O \rightarrow [Co(AA)_2(H_2O)Cl]^{2+} + Cl^-$$

AA	10^3k, min^{-1}
NH_2—CH_2—CH_2—NH_2(en)	1·9
NH_2—CH_2—$CH(CH_3)$—NH_2(pn)	3·7
dl-NH_2—$CH(CH_3)$—$CH(CH_3)$—NH_2(*dl*-bn)	8·8
meso-NH_2—$CH(CH_3)$—$CH(CH_3)$—NH_2(*m*-bn)	250
NH_2—$C(CH_3)_2$—$C(CH_3)_2$—NH_2(tetrameen)	very fast
NH_2—CH_2—CH_2—CH_2—NH_2(tn)	600 (10°)

* From ref. (4).

An alternative explanation which suggests itself is that the rate increase may be due to inductive effects: increasing alkyl substitution will distort the electron density towards the cobalt atom, assisting the dissociation of the chloride ion. An examination of the influence of such effects, uncomplicated by steric factors, has been made[5] on a series of substituted pyridine complexes of the type $[Co(en)_2(X\text{-}py)Cl]^{2+}$ (Table 2.5). That an increasing inductive effect

Table 2.5

Rate constants for acid hydrolysis of $[Co(en)_2(X\text{-}py)Cl]^{2+}$, at 50°*

X-py = *meta*- or *para*-substituted pyridine

X-py	pK_B	10^5k, s^{-1}
pyridine	8·82	1·1
3-methylpyridine	8·19	1·3
4-methylpyridine	7·92	1·4
4-methoxypyridine	7·53	1·5

* From ref. (5).

is operating in this sequence is evidenced by the increasing basic strength of the pyridine molecule.* It is seen from the rate and base

* A base is an electron-pair donor *and* a proton acceptor.

dissociation constants that a 20-fold increase in basic strength leads to a 40 per cent increase in rate. However, the basic strengths of the substituted ethylenediamines in Table 2.4 vary only by a factor of about 1·5 and thus it can be concluded with confidence that the very marked rate increase is due to other influences than inductive effects.

In the acid hydrolysis of pentaminechloro- and tetraminedichloro-Co(III) complexes[6] the effects of charge and degree of chelation are evident (Table 2.6). As the overall charge on the complex increases

Table 2.6

Rate constants for acid hydrolysis of some pentaminechloro- (35°) and tetraminedichloro- (25°) complexes of Co(III) at pH1*

Complex†	10^7k, s^{-1}	Complex	10^7k, s^{-1}
cis-$[Co(NH_3)_4Cl_2]^+$	very fast	$[Co(NH_3)_5Cl]^{2+}$	67
cis-$[Co(en)_2Cl_2]^+$	2400	*cis*-$[Co(en)_2NH_3Cl]^{2+}$	14
cis-$[Co(trien)Cl_2]^+$	1500	*cis*-$[Co(trien)NH_3Cl]^{2+}$	6·7
trans-$[Co(NH_3)_4Cl_2]^+$	18,000	$[Co(en)(dien)Cl]^{2+}$	5·2
trans-$[Co(en)_2Cl_2]^+$	350	$[Co(tetraen)Cl]^{2+}$	2·5

* From ref. (6).
† Substitution of first Cl^- only for dichloro complexes.

the rate of substitution decreases, which is indicative of a dissociative mechanism (cf. Table 2.1). The marked reduction in rate with increasing degrees of chelation, e.g., in the series of monochloro-complexes, is more difficult to interpret.

The most acceptable explanation is in terms of a solvent effect. A dissociation reaction proceeds via a transition state by separating charges, and such separation of charge is aided the greater the solvation of the activated complex. Thus factors contributing to poorer solvation (and hence destabilization) of the transition state lead to lower rates of reaction. On increased chelation the solvation shell of the complex becomes more and more disturbed by the bulky organic ligands, resulting in a greater energy barrier for reaction and a corresponding reduction in rate.

It is important to examine the influence of non-replaceable anionic ligands by a study of the hydrolysis of a series of closely related complexes. A suitable example is found in the effect of changes of ligand

L on the rate of aquation of $[Co(en)_2LCl]^{n+}$. A reasonable rationalization of the trends shown in Table 2.7 is in terms of electronic effects

Table 2.7

Rate constants and steric course of reaction*

$$[Co(en)_2LCl]^{n+} + H_2O \xrightarrow{25^\circ} [Co(en)_2L(H_2O)]^{(n+1)+} + Cl^-$$

Isomer	Ligand (L)	10^7k, s^{-1}	% *cis* product
cis	OH^-	130,000	100
cis	N_3^-	2400	100
cis	Cl^-	2400	100
cis	SCN^-	110	100
cis	NO_2^-	1100	100
trans	OH^-	14,000	75
trans	N_3^-	2500	20
trans	Cl^-	350	35
trans	SCN^-	0·5	50–70
trans	NO_2^-	9800	0

* From refs (6a) and (6c).

of the substituents.[6] The substituents used in these investigations are of two types: (*a*) ligands possessing lone pairs of electrons which can release electrons to the metal by an inductive process:

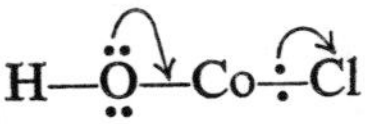

H—Ö—Co—Cl

That is, delocalization of the electron pair, M $\leftleftharpoons$ L, occurs, resulting in some π-bonding. Examples of this class of ligand are OH^- and Cl^-. (*b*) Ligands (e.g., NO_2^- and CN^-) which can withdraw electrons from the metal centre by delocalization of electron pairs towards the ligand, M $\rightleftharpoons$ L:

O_2N—Co—Cl

With these distinctions in mind it becomes possible to forecast the effect of π-bonding on the mechanism of the aquation of $[Co(en)_2LCl]^{n+}$ for ligand types (*a*) and (*b*). For type (*a*), electron displacement occurs from the ligand to the metal making it easier to break the Co—Cl bond. Also such π-bonding effectively stabilizes the five-coordinated intermediate. A dissociative process is supported by both effects. For type (*b*) ligands, π-bonding in the direction metal to ligand results in the strengthening of the Co—Cl bond, but at the same time invites nucleophilic attack of solvent by an associative mechanism:

$$\mathrm{H\ddot{O}{-}Co{-}Cl} \longrightarrow \mathrm{H\overset{\oplus}{O}{=}Co + Cl^-}$$

$$\mathrm{O_2N{-}Co{-}Cl} \longrightarrow \mathrm{O{=}N(O^{\ominus}){=}\overset{\oplus}{Co}\cdots Cl,\ \cdots OH_2}$$

Although the assignment of a dissociative mechanism seems reasonable enough for π-donors like OH^-, an immediate difficulty is met with in the case of π-acceptors like NO_2^-. For the latter group, the proposed transition state implies that the formation of the Co—OH_2 bond takes precedence over the fission of the Co—Cl bond, since new bond formation will be easier than breakage of an existing stronger bond. It can be concluded therefore that the nature of the entering group should affect the rate of hydrolysis. However, no effects of this type have been detected; the rates of solvolysis of $[Co(en)_2NO_2Cl]^+$ by dimethylformamide, methanol, and dimethylsulphoxide do not differ radically from the rate of aquation. It seems more likely, therefore, that the function of the NO_2^- or CN^- groups is not anomalous, and that the hydrolysis of $[Co(en)_2LCl]^+$ proceeds via a dissociative mechanism independent of the nature of the non-labile ligand L. In support there is definite evidence to suggest that π-acceptors *can* stabilize intermediates of *reduced* coordination number, e.g., the well-documented five-coordinated intermediate, $[Co(CN)_5]^{2-}$ (see p. 38).

Some interesting considerations of steric factors have arisen from the kinetic results of $[Co(en)_2LCl]^+$ hydrolysis. For the ligands OH^-, Cl^-, and NCS^-, it is seen from Table 2.7 that the *cis* isomers react more rapidly than the *trans*. Also the *cis* isomers, unlike the *trans*, react with retention of configuration. A likely explanation, in terms of π-bonding, has been advanced. The expulsion of Cl^-, and hence the dissociation of *cis*-$[CoA_4LCl]^+$, via a five-coordinated, tetragonal activated complex will take place readily as a *p*-orbital on the non-labile ligand L, e.g., OH^-, can π-bond with ease to a vacated *p*-orbital or d^2sp^3 hybrid orbital of Co(III) (see Fig. 2.1). Also on this scheme,

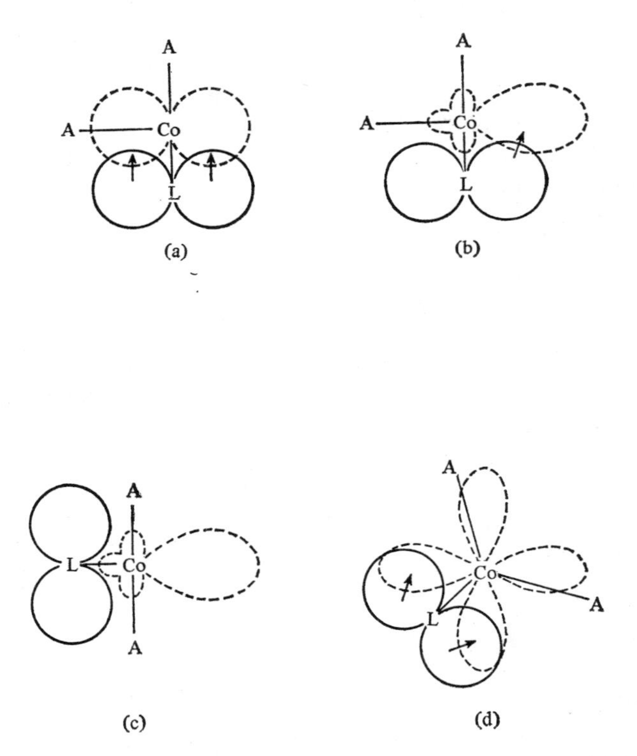

Fig. 2.1. Overlap of filled *p* orbital of OH^-(L) with vacant (*a*) *p*-orbital, or (*b*) d^2sp^3 hybrid orbital of Co(III) in a five-coordinated tetragonal pyramid activated complex resulting from the dissociation of Cl^- from *cis*-$[CoA_4OHCl]^+$. (*c*) No overlap of filled *p*-orbital of OH^- with vacant d^2sp^3 hybrid orbital of Co(III) in the tetragonal pyramid resulting from the dissociation of Cl^- from *trans*-$[CoA_4OHCl]^+$. (*d*) Efficient overlap with vacant $d_{x^2-y^2}$ orbital after rearrangement to a trigonal bipyramid structure The two A ligands not shown are above and below the plane of the paper. From F. Basolo and R. G. Pearson, in *Progress in Inorganic Chemistry* (ed. F. A. Cotton), Vol. 4, p. 448, Interscience, 1962.

as no re-arrangement is necessary, *cis* isomers should react with retention of configuration. In the case of *trans*-$[CoA_4LCl]^+$ the *p*-orbital on OH^- cannot overlap a vacated d^2sp^3 orbital without substantial rearrangement. The rate of aquation of such *trans* isomers is thus predicted to be lower than that of the analogous *cis* isomers. Rearrangement to a trigonal bipyramid structure has been shown to leave a vacant $d_{x^2-y^2}$ orbital in the trigonal plane which can accept electrons by π-bonding from the OH^-. Consideration of the stereochemistry of this mechanism reveals that a mixture of *cis* and *trans* products will result on aquation. By way of contrast, the NO_2^- group has the property of increasing hydrolysis of $[Co(en)_2NO_2Cl]^+$ more in the *trans* position than in the *cis*, both isomers reacting with retention of configuration. This *trans* effect is the norm for square planar complexes of Pt(II) (cf. p. 67).

Acid-catalysed aquation

The discussion until now has been concerned with systems where hydrolysis takes place under acid conditions but is not dependent on the concentration of acid, provided this is less than pH 4. For some systems, however, the rate of hydrolysis is acid-catalysed and a protonated species is formed as a reactive intermediate. These systems fall into two categories: (*a*) complexes containing ligands derived from weak acids, e.g., $[M(NH_3)_5X]^{n+}$ where $X = F^-$, N_3^-, $RCOO^-$, CO_3^{2-} or NO_2^-, and $[Coen_2X_2]^+$ where $X = NO_2^-$ or N_3^-; (*b*) complexes containing polydentate ligands, e.g., $[M(EDTA)]^{n-}$ where M = Ni(II) or Fe(III), $[M(C_2O_4)_3]^{3-}$ where M = Co(III), Cr(III) or Rh(III), $[M(bipy)_3]^{2+}$ where M = Fe(II) or Ni(II), and $[Ni(en)_3]^{2+}$.

For complexes in category (*a*), acid catalysis is due to the tendency of the strongly basic ligand to attach a proton,[7] for example

$$[M(NH_3)_5(RCOO)]^{2+} + H^+ \overset{K}{\rightleftharpoons} [M(NH_3)_5(RCOOH)]^{3+} \quad \text{fast}$$

Aquation then results through the alternative paths

$$[M(NH_3)_5(RCOO)]^{2+} + H_2O \xrightarrow{k_1} [M(NH_3)_5(H_2O)]^{3+} + RCOO^-$$

$$[M(NH_3)_5(RCOOH)]^{3+} + H_2O \xrightarrow{k_2} [M(NH_3)_5(H_2O)]^{3+} + RCOOH$$

Since the concentration of protonated species is small even under strongly acid conditions, the rate law contains two terms

$$\text{rate} = k_1[M(NH_3)_5(RCOO)^{2+}] + k_{H^+}[M(NH_3)_5(RCOO)^{2+}][H^+]$$

where $k_{H^+} = k_2K$. The observed rate constant (k_{obs}) is then defined by $k_{obs} = k_1 + k_2K[H^+]$, and a plot of k_{obs} versus $[H^+]$ is linear with an intercept equal to k_1 and a gradient of k_2K. This is illustrated by Fig. 2.2 for a series of pentammine complexes of Co(III), Rh(III), and

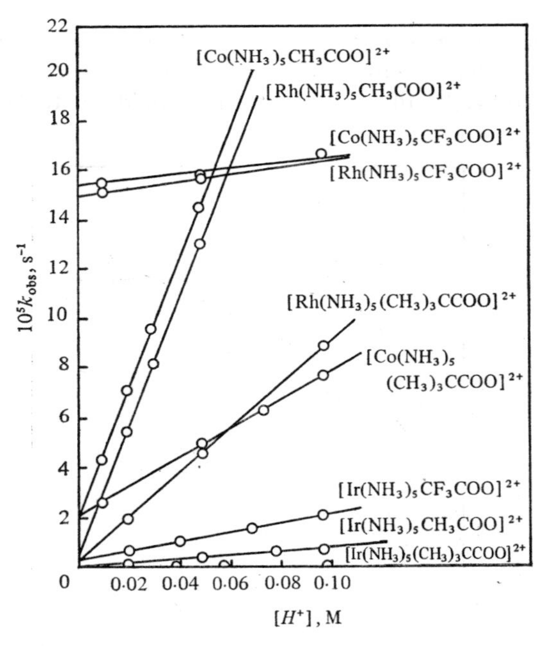

Fig. 2.2. Rates of hydrolysis of $[M(NH_3)_5RCOO]^{2+}$ versus $[H^+]$ at 80° and $\mu = 0{\cdot}10$ M. From F. Monacelli, F. Basolo and R. G. Pearson, *J. Inorg. Nucl. Chem.*, 1962, **24**, 1241.

Ir(III). A variation on the above mechanism[8] occurs in the acid-catalysed aquation of azidopentacyanocobaltate(III). In the absence of SCN^- the reaction proceeds by

$$[Co(CN)_5N_3]^{3-} + H^+ \underset{}{\overset{K}{\rightleftharpoons}} [Co(CN)_5N_3H]^{2-} \qquad \text{fast}$$

$$[Co(CN)_5N_3H]^{2-} + H_2O \xrightarrow{\text{slow}} [Co(CN)_5OH_2]^{2-} + HN_3$$

but in the presence of SCN^- aquation is thought to occur through the pentacyanocobaltate(III) intermediate

$$[Co(CN)_5N_3H]^{2-} \xrightarrow{\text{slow}} [Co(CN)_5]^{2-} + HN_3$$

$$[Co(CN)_5]^{2-} + H_2O \rightarrow [Co(CN)_5OH_2]^{2-}$$

in competition with

$$[Co(CN)_5]^{2-} + SCN^- \rightarrow [Co(CN)_5NCS]^{3-}$$

Conceivably a similar mechanism could operate for complexes containing multidentate ligands (category (*b*) above), e.g., in the dissociation of $[Fe(bipy)_3]^{2+}$ the acid-dependence could be attributed to the

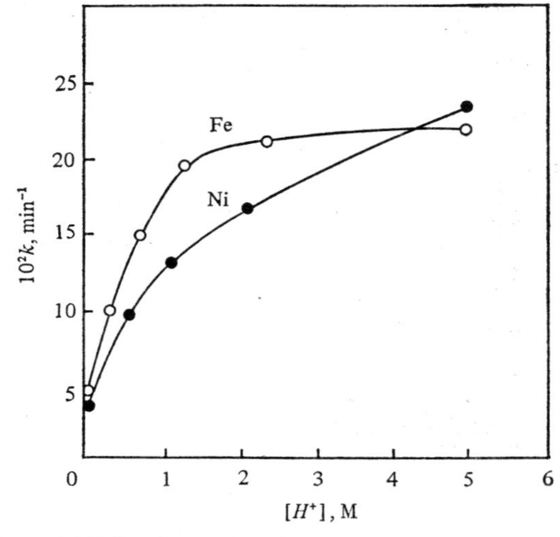

Fig. 2.3. Effect of $[H^+]$ on rate of dissociation of $[Fe(bipy)_3]^{2+}$ (at 25°) and $[Ni(bipy)_3]^{2+}$ (at 13°). From F. Basolo, in *Survey of Progress in Chemistry* (ed. A. F. Scott), Vol. 2, p. 27, Academic Press, 1964.

formation of a protonated intermediate $[Fe(bipy)_3H]^{3+}$. It would be expected, however, that the weak basicity of the bipyridyl group would not favour appreciable concentrations of such entities. On this basis also it is difficult to explain the limiting rate obtained at high acid concentration (Fig. 2.3) which implies that protonation is complete. A more probable explanation[9] is in terms of a mechanism in which an opening-up of the chelate ring occurs, followed by protonation of the free end and then loss of the protonated ligand:

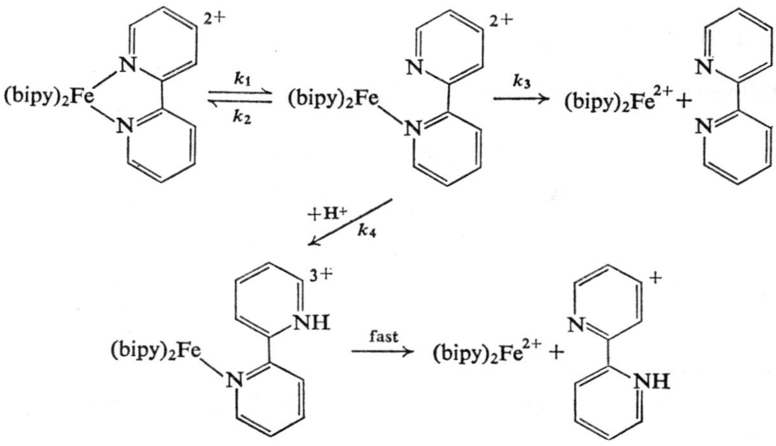

(bipy)$_2$Fe(bipy)$^{2+}$ $\underset{k_2}{\overset{k_1}{\rightleftharpoons}}$ (bipy)$_2$Fe–N(bipy, one end free)$^{2+}$ $\xrightarrow{k_3}$ (bipy)$_2Fe^{2+}$ + bipy

(bipy)$_2$Fe–N(bipy, one end free)$^{2+}$ $\xrightarrow[k_4]{+H^+}$ (bipy)$_2$Fe–N(bipy, NH)$^{3+}$ $\xrightarrow{\text{fast}}$ (bipy)$_2Fe^{2+}$ + bipyH^{+}

Application of the stationary-state approximation to the concentrations of the intermediate species gives the observed rate constant as

$$k_{obs} = \frac{k_1(k_3 + k_4[H^+])}{k_2 + k_3 + k_4[H^+]}$$

Two limiting cases arise: at very low hydrogen-ion concentrations, $k_{obs} = k_1k_3/(k_2 + k_3)$; at high $[H^+]$, $k_4[H^+] \gg (k_2 + k_3)$ and $k_{obs} = k_1$. An approximate value of 0·2 for $k_3/(k_2 + k_3)$ was obtained from the ratio of these two limiting rates. This result indicates that, on the average, 80 per cent of the occasions on which the first metal–nitrogen bond breaks result in re-formation of the bond, and 20 per cent lead to a second bond being broken and dissociation taking place. A similar type of mechanism to the one above is applicable to the dissociation of $[Ni(en)_3]^{2+}$ and $[Ni(bipy)_3]^{2+}$.

In the case of the acid-catalysed aquation of trioxalatochromate(III)

$$[Cr(C_2O_4)_3]^{3-} + 2H_3O^+ \rightarrow [Cr(H_2O)_2(C_2O_4)_2]^- + H_2C_2O_4$$

the rate law assumes the form

$$-d[Cr(C_2O_4)_3{}^{3-}]/dt = k'[H_3O^+][Cr(C_2O_4)_3{}^{3-}] + k''[H_3O^+]^2[Cr(C_2O_4)_3{}^{3-}]$$

The mixed-order dependence on the hydrogen-ion concentration is shown [10] to be consistent with a rapid pre-equilibrium protonation of the complex

$$[Cr(C_2O_4)_3]^{3-} + H_3O^+ \overset{K}{\rightleftharpoons} [Cr(H_2O)(C_2O_4)_2(HC_2O_4)]^{2-}$$

followed by the parallel rate-determining steps

$$[Cr(H_2O)(C_2O_4)_2(HC_2O_4)]^{2-} + H_2O \xrightarrow{k_1} [Cr(H_2O)_2(C_2O_4)_2]^- + HC_2O_4^-$$

and

$$[Cr(H_2O)(C_2O_4)_2(HC_2O_4)]^{2-} + H_3O^+ \xrightarrow{k_2} [Cr(H_2O)_2(C_2O_4)_2]^- + H_2C_2O_4$$

where $k' = k_1K$ and $k'' = k_2K$.

Finally, it should be added that in some reactions an inverse acid-dependence is noted, e.g., in the aquation of *trans*-dichlorotetra-aquochromium(III)

$$[CrCl_2(H_2O)_4]^+ + H_2O \rightarrow [CrCl(H_2O)_5]^{2+} + Cl^-$$

The suggested mechanism [11]

$$[CrCl_2(H_2O)_4]^+ \overset{\text{fast}}{\rightleftharpoons} [CrCl_2(H_2O)_3OH] + H^+$$

$$[CrCl_2(H_2O)_3OH] + H_2O \xrightarrow{\text{slow}} [CrCl(H_2O)_4OH]^+ + Cl^-$$

$$H^+ + [CrCl(H_2O)_4OH]^+ \overset{\text{fast}}{\rightleftharpoons} [CrCl(H_2O)_5]^{2+}$$

is in agreement with the empirical rate law, $k_{obs} = k' + k''/[H^+]$, where the first term represents the direct rate of aquation of *trans*-$[CrCl_2(H_2O)_4]^+$.

Anation reactions

The replacement of water from an aquo complex by an anionic group (the reverse of aquation) is called *anation*, for example,

$$[CoA_5H_2O]^{3+} + Y^- \rightarrow [CoA_5Y]^{2+} + H_2O$$

The fact that some anation reactions conform to a second-order rate law does not necessarily indicate an associative mechanism. The following dissociative scheme is more likely:

$$[CoA_5H_2O]^{3+} \underset{k_2}{\overset{k_1}{\rightleftharpoons}} [CoA_5]^{3+} + H_2O$$

$$[CoA_5]^{3+} + Y^- \xrightarrow{k_3} [CoA_5Y]^{2+}$$

$$\text{rate} = d[CoA_5Y^{2+}]/dt = k_3[CoA_5{}^{3+}][Y^-]$$

By assuming steady-state conditions to apply

$$[CoA_5{}^{3+}] = \frac{k_1[CoA_5H_2O^{3+}]}{k_2 + k_3[Y^-]}$$

and hence

$$\frac{d[CoA_5Y^{2+}]}{dt} = \frac{k_1k_3[CoA_5H_2O^{3+}][Y^-]}{k_2 + k_3[Y^-]}$$

If $[Y^-]$ is low and $k_2 \gg k_3[Y^-]$,

$$\text{then rate} = (k_1k_3/k_2)[CoA_5H_2O^{3+}][Y^-]$$

For high $[Y^-]$ and $k_3[Y^-] \gg k_2$,

$$\text{then rate} = k_1[CoA_5H_2O^{3+}]$$

Depending on the conditions applying, the rate may or may not be dependent on the concentration of the entering substituent. In principle, a gradual change from second- to first-order kinetics may be revealed as the concentration of substituent is increased. Once this limit is reached the rate should then be independent of the nature of the incoming group, and the observed rate constant should be equal to that for isotopic exchange between water and aquo complex.

Attempts to detect these effects, in reactions of cobalt(III) ammines with azide, thiocyanate, and sulphate, were frustrated by outer-sphere association between the oppositely-charged reactants at the high anionic concentrations required to achieve the anion-independent limit. Under these conditions an ion-pair path, which is kinetically second order, competes against reaction via a dissociative process:

$$[CoA_5H_2O]^{3+} + Y^- \underset{k_2}{\overset{k_1}{\rightleftharpoons}} [CoA_5H_2O]^{3+}.Y^- \qquad \text{(ion-pair)}$$

$$[CoA_5H_2O]^{3+}.Y^- \xrightarrow{k_3} [CoA_5Y]^{2+} + H_2O$$

where $$\frac{d[CoA_5Y^{2+}]}{dt} = \frac{k_1k_3}{k_2+k_3}[CoA_5H_2O^{3+}][Y^-]$$

Avoiding the complications of ion-pairing by the simple expedient of using an anionic rather than a cationic complex of Co(III), Haim and Wilmarth[12] were successful in demonstrating a dissociative mechanism in the reaction of aquopentacyanocobaltate(III), $[Co(CN)_5H_2O]^{2-}$, with N_3^- and SCN^- ions. Their kinetic results are consistent with the following scheme:

$$[Co(CN)_5H_2O]^{2-} \underset{k_2}{\overset{k_1}{\rightleftharpoons}} [Co(CN)_5]^{2-} + H_2O$$

$$[Co(CN)_5]^{2-} + Y^- \xrightarrow{k_3} [Co(CN)_5Y]^{3-}$$

At high Y^- concentrations the reaction becomes pseudo-first order in $[Co(CN)_5H_2O]^{2-}$:

$$d[Co(CN)_5Y^{3-}]/dt = k_{obs}[Co(CN)_5H_2O^{2-}]$$

where $$k_{obs} = \frac{k_1k_3[Y^-]}{k_2+k_3[Y^-]}$$

On rearrangement of this equation

$$\frac{1}{k_{obs}} = \frac{k_2}{k_1k_3[Y^-]} + \frac{1}{k_1}$$

and values of k_1 and k_2/k_3 can be obtained from the intercept and slope–intercept ratio of a plot of $1/k_{obs}$ versus $1/[Y^-]$ for varying $[Y^-]$. The values of rate constants are listed in Table 2.8. It is seen

Table 2.8

Rate constants* for the reaction:
$[Co(CN)_5H_2O]^{2-} + Y^- \rightarrow [Co(CN)_5Y]^{3-} + H_2O$

Entering group	40°, $\mu = 1$ M		20°, $\mu = 5{\cdot}0$ M	
	10^3k_1, s^{-1}	k_2/k_3	10^4k_1, s^{-1}	k_2/k_3
N_3^-	1·4	1·9	4·7	3·0
SCN^-	1·8	2·95	5·4	5·0

* From ref. (12).

that the reasonable constancy of the k_1 values for N_3^- and SCN^-, under the two sets of conditions, is a convincing test of the above mechanism. Further support is given by the rate of water exchange for $[Co(CN)_5OH_2]^{2-}$ with $H_2{}^{18}O$ which was found to have a rate constant between $1{\cdot}0 \times 10^{-3}$ and $1{\cdot}3 \times 10^{-3}$ s^{-1} for $\mu = 1$ M and 40°. The values of k_2/k_3 reveal that azide is a better nucleophile to $[Co(CN)_5]^{2-}$ than thiocyanate. In an extension of this work[13] to a number of other ligands the following order of reactivity was revealed:

$$OH^- > N_3^- > SCN^- > I^- > NH_3 > Br^- > S_2O_3{}^{2-} > CNO^- > H_2O$$

In this series, the position of water is that calculated for a concentration equal to that of the other ligands. It is interesting to consider whether a dissociative mechanism might have been anticipated. In organic systems a highly electronegative group like CN^- favours an associative (S_N2) rather than a dissociative process, but in the case of $[Co(CN)_5OH_2]^{2-}$ the five cyanide groups would give rise to a high electron density at the cobalt centre, resulting in a relatively weak Co—OH_2 bond. Also an increase of bond angle would be expected to stabilize the five-coordinated intermediate $[Co(CN)_5]^{2-}$.

An analogous mechanism has been confirmed recently[14] in a study of some replacement reactions of substituted sulphitopentammine-cobalt(III) complexes:

$$[Co(NH_3)_4SO_3(X)] + Y \rightarrow [Co(NH_3)_4SO_3(Y)] + X$$

where (*a*) $X = SCN^-$, $Y = NH_3$; $X = NO_2^-$, $Y = NH_3$; (*b*) $X = OH^-$, $Y = NH_3$; (*c*) $X = NH_3$, $Y = OH^-$, CN^-, NO_2^- or SCN^-; and (*d*) $X = OH^-$, $Y = CN^-$. These processes apparently proceed via a common dissociative mechanism which is

$$[Co(NH_3)_4SO_3(X)] \underset{k_2}{\overset{k_1}{\rightleftharpoons}} [Co(NH_3)_4SO_3]^+ + X$$

$$[Co(NH_3)_4SO_3]^+ + Y \xrightarrow{k_3} [Co(NH_3)_4SO_3(Y)]$$

with a rate law given by

$$\frac{-d[Co(NH_3)_4SO_3(X)]}{dt} = \frac{k_1k_3[Co(NH_3)_4SO_3(X)][Y]}{k_2[X] + k_3[Y]}$$

The reactions were followed spectrophotometrically at 25° and at a constant ionic strength of $\mu = 0{\cdot}46$ M, using a stopped-flow apparatus for fast rates. For reactions in group (a) above, the full form of rate law applied over the whole concentration range studied of replacing and replaced ligand. For reaction (*b*) a full form of rate law changed over to a limiting (entering-group independent) form. For reactions in (*c*) only the limiting form was observed with

$$\text{rate} = k_1[Co(NH_3)_4SO_3(NH_3)^+]$$

Reaction (*d*) was unique in adhering to a rate law given by

$$\text{rate} = k_1k_3[Co(NH_3)_4SO_3(OH)][CN^-]/k_2[OH^-]$$

which corresponds to the second limiting condition of the full rate law if $k_2[X] \gg k_3[Y]$. From the kinetic data on the reactions in groups (*a*) and (*b*), values were obtained for the dissociation rate constants (k_1^X) of the various complexes $[Co(NH_3)_4SO_3(X)]$ and the relative reactivities of the various nucleophiles ($k_2^X/k_2^{NH_3}$) towards the intermediate $[Co(NH_3)_4SO_3]^+$ (Table 2.9).

Table 2.9

Rate constants* for the reaction:
$[Co(NH_3)_4SO_3X] + NH_3 \rightarrow [Co(NH_3)_4SO_3(NH_3)] + X$

X	NH_3	NO_2^-	SCN^-	OH^-
k_1^X, s^{-1}	0·012	0·46	1·75	7
$k_2^X/k_2^{NH_3}$	1	70	30	8000

* From ref. (14).

Base hydrolysis

For octahedral complexes there is little evidence to suggest that substitution of one ligand by another takes place directly. Instead, as mentioned earlier, it appears that replacement occurs by the prior formation of an aquo complex, as, for instance:

$$[Co(en)_2(NO_2)Cl]^+ + H_2O \overset{\text{slow}}{\rightleftharpoons} [Co(en)_2(NO_2)(H_2O)]^{2+} + Cl^-$$

$$[Co(en)_2(NO_2)(H_2O)]^{2+} + Y^- \xrightarrow{\text{fast}} [Co(en)_2(NO_2)Y]^+ + H_2O$$

In this example, the rate of chloride-ion release is independent of the concentration of the incoming ligand (Y) for $Y = N_3^-$, NO_2^-, or

SCN^-. Likewise the rate of chloride-ion exchange of *cis*-$[Co(en)_2Cl_2]^+$ is independent of the concentration of chloride ion

$$[Co(en)_2Cl_2]^+ + H_2O \xrightarrow{\text{slow}} [Co(en)_2Cl(H_2O)]^{2+} + Cl^-$$

$$[Co(en)_2Cl(H_2O)]^{2+} + \overset{*}{Cl}{}^- \xrightarrow{\text{fast}} [Co(en)_2Cl\overset{*}{Cl}]^+ + H_2O$$

The exchange reactions of $\overset{*}{N}CS^-$ with $[Co(en)_2(NCS)_2]^+$, $[Co(NH_3)_5NCS]^{2+}$, and $[Cr(NH_3)_5NCS]^{2+}$ show similar behaviour.

In base hydrolysis, however, hydroxide ions have the special property of replacing coordinated ligands without the intervention

Table 2.10

Rate constants for acid hydrolysis (k_{H_2O}) and base hydrolysis (k_{OH^-})*
for some Co(III) complexes, at 25°

	k_{H_2O}, s^{-1}	k_{OH^-}, M^{-1} s^{-1}
$[Co(NH_3)_5Cl]^{2+}$	$1{\cdot}7 \times 10^{-6}$	$8{\cdot}5 \times 10^{-1}$
cis-$[Co(en)_2Cl_2]^+$	$2{\cdot}5 \times 10^{-4}$	1×10^{3}
trans-$[Co(NH_3)_4Cl_2]^+$	$1{\cdot}8 \times 10^{-2}$	$1{\cdot}8 \times 10^{3}$
trans-$[Co(en)_2Cl_2]^+$	$3{\cdot}2 \times 10^{-5}$	$3{\cdot}0 \times 10^{3}$
trans-$[Co(en)_2Br_2]^+$	$1{\cdot}4 \times 10^{-4}$	$1{\cdot}2 \times 10^{4}$
trans-$[Co(en)_2F_2]^+$	1×10^{-6}	$6{\cdot}4 \times 10^{1}$
cis-$[Co(trien)Cl_2]^+$	$1{\cdot}5 \times 10^{-4}$	2×10^{5}

* From ref. (15).

of aquo complexes. In this respect, the role of the hydroxide ion is a unique one in that it is the only nucleophile which can successfully compete with water. The full form of rate law observed for complexes subject to both acid and base hydrolysis is

$$-\mathrm{d}[complex]/\mathrm{d}t = k_{H_2O}[complex] + k_{OH^-}[OH^-][complex]$$

and a comparison of the values[15] of rate constants for some halogenoaminecobalt(III) complexes listed in Table 2.10 shows the rate of base hydrolysis (at, say, pH 13) to be some 10^4 to 10^7 times greater than that of aquation.

Using $[Co(NH_3)_5Cl]^{2+}$ as an example, the most obvious mechanism to account for the first-order dependence on OH^- is an associative (S_N2) one[16] in which the OH^- ion directly replaces the halide group X

$$[Co(NH_3)_5X]^{2+} + OH^- \rightarrow [Co(NH_3)_5OH]^{2+} + X^-$$

An alternative to this is a scheme[17] involving a rapidly established pre-equilibrium in which the hydroxide ion forms an amido conjugate base by abstracting a proton from one of the ammine nitrogens. The rate-determining step is the dissociation of the conjugate base to give a penta-coordinated intermediate which then adds solvent to form the hydroxo product:

$$[Co(NH_3)_5X]^{2+} + OH^- \overset{K}{\rightleftharpoons} [Co(NH_3)_4NH_2X]^{+} + H_2O \qquad \text{fast}$$

$$[Co(NH_3)_4NH_2X]^{+} \xrightarrow{k} [Co(NH_3)_4NH_2]^{2+} + X^- \qquad \text{rate-determining}$$

$$[Co(NH_3)_4NH_2]^{2+} + H_2O \rightarrow [Co(NH_3)_5OH]^{2+} \qquad \text{fast}$$

This mechanism has become known by the symbol S_N1CB (where CB = conjugate base). The rate follows as

$$\begin{aligned} -\mathrm{d}[Co(NH_3)_5X^{2+}]/\mathrm{d}t &= k[Co(NH_3)_4NH_2X^{+}] \\ &= kK[Co(NH_3)_5X^{2+}][OH^-] \\ &= \frac{kK_A}{K_W}[Co(NH_3)_5X^{2+}][OH^-] \end{aligned}$$

where K_A is the acid dissociation constant of the complex and K_W is the ionic product of water. In keeping with the empirical rate law, it is required that establishment of the pre-equilibrium is very rapid in comparison with the rate of the overall process and, also, that the concentration of the amido conjugate-base should be low (otherwise a limiting rate would be attained at a sufficiently high pH). Seemingly, both requirements are fulfilled. Firstly, exchange studies of $[Co(NH_3)_5Cl]^{2+}$ with D_2O in basic solution have shown that the exchange rate is faster than the release of chloride ion. Secondly, cobalt(III) amine complexes are such weak acids that the concentration of base form is very low, even in strongly alkaline solution. Indirect evidence in support of the mechanism is plentiful:

(*a*) Increasing chelation is paralleled by an increase in the rate of base hydrolysis (see Table 2.10). The cobalt(III) amine complexes are so weakly acidic that their acid strengths have not been measured but it is known that in the case of the analogous platinum(IV) complexes increased chelation gives rise to increased acidity. Also complexes which do not possess acidic protons like *trans*-$[Co(py)_4Cl_2]^+$ and $[Co(CN)_5X]^{3-}$ undergo slow base hydrolysis at rates independent of hydroxide ion concentration.

(*b*) The isotopic ratio of ^{18}O to ^{16}O in water is known not to be the same as the ratio in the hydroxide ion, although rapid proton transfer occurs between the ion and the solvent. This, in effect, labels the hydroxide ion. For the base hydrolysis of $[Co(NH_3)_5X]^{2+}$, Taube and Green[18] showed that the final hydroxo product, $[Co(NH_3)_5OH]^{2+}$, must derive from the reaction of the intermediate, $[Co(NH_3)_4NH_2]^{2+}$, with water and *not* with hydroxide ion, since the ^{18}O–^{16}O ratio in the product was identical with the ratio in water.

(*c*) In the base hydrolysis of complexes of the type $[Co(en)_2AX]^{n+}$, where A remains constant and X (the outgoing ligand) is varied, the concentration of the isomeric products should be independent of the nature of X if a S_N1CB mechanism applies, since reaction occurs via a common intermediate.[19] Some representative data are given in Table 2.11, and show the remarkable agreement between the proportion of *cis*-product from different parents.[20]

Detection of the conjugate-base intermediate would constitute direct proof of the mechanism. Although some evidence exists for this five-coordinated species in dimethylsulphoxide solvent (see p. 50), attempts, using spectrophotometry, to detect a similar intermediate in aqueous solution were unsuccessful due to the formation of ion-pairs.

This suggests a third possible mechanism for base hydrolysis which has been applied[21] to Co(III)-ethylenediamine complexes of the type $[Co(en)_2LCl]^{2+}$ where L, for example, is hydroxylamine. For these systems an interesting result emerges: the rate of base hydrolysis is less for the chloro-hydroxylamine complex than for the corresponding chloro-ammine one, despite the fact that hydroxylamine is more acidic than ammonia. Apparently in this case, although not necessarily in general, a conjugate-base process does not apply. The explanation offered instead is in terms of an ion-pair mechanism:

$$[Co(en)_2LCl]^{2+} + OH^- \underset{}{\overset{K}{\rightleftharpoons}} [Co(en)_2LCl]^{2+}.OH^- \qquad \text{(ion-pair)}$$

$$[Co(en)_2LCl]^{2+}.OH^- \xrightarrow{k} [Co(en)_2LOH]^{2+} + Cl^-$$

Table 2.11

Isomeric products of base hydrolysis* of *cis*- and *trans*- $[Co(en)_2AX]^{n+}$

A = non-replaced ligand
X = replaced ligand

Reactant	Product	% *cis*-isomer in product	
		For *cis*- reactant	For *trans*- reactant
$[Co(en)_2ClBr]^+$	$[Co(en)_2ClOH]^+$	30	5
$[Co(en)_2ClCl]^+$	$[Co(en)_2ClOH]^+$	37	5
$[Co(en)_2(NCS)N_3]^+$	$[Co(en)_2(NCS)OH]^+$	70	70
$[Co(en)_2(NCS)Cl]^+$	$[Co(en)_2(NCS)OH]^+$	80	76
$[Co(en)_2(NCS)Br]^+$	$[Co(en)_2(NCS)OH]^+$	—	81
$[Co(en)_2(OH)Br]^+$	$[Co(en)_2(OH)OH]^+$	96	90
$[Co(en)_2(OH)Cl]^+$	$[Co(en)_2(OH)OH]^+$	97	94
$[Co(en)_2(NO_2)Cl]^+$	$[Co(en)_2(NO_2)OH]^+$	67	6
$[Co(en)_2(NO_2)NCS]^+$	$[Co(en)_2(NO_2)OH]^+$	55	10
$[Co(en)_2(N_3)Cl]^+$	$[Co(en)_2(N_3)OH]^+$	59	27
$[Co(en)_2(N_3)N_3]^+$	$[Co(en)_2(N_3)OH]^+$	55	30

* From ref. (19), and references cited therein.

This interchange scheme, a subtle variation on collision-activated bimolecular substitution, avoids the postulation of an intermediate of altered coordination number. To be compatible with the observed second-order kinetics it is necessary that the equilibrium concentration of the ion-pair is low. The rate law is then

$$\text{rate} = k[\textit{ion-pair}] = kK[\textit{complex}][OH^-]$$

More precisely, ion-pairing of this type can be regarded[22] in terms of solvation changes of the complex ion, in which the hydroxide ion replaces a solvent molecule in the inner solvation shell:

$$\{[\text{Coen}_2\text{LCl}](\text{H}_2\text{O})_n\}^{2+} + \text{OH}^- \overset{K}{\rightleftharpoons} \{[\text{Coen}_2\text{LCl}]\text{OH}(\text{H}_2\text{O})_{n-1}\}^+ + \text{H}_2\text{O}$$

$$\{[\text{Coen}_2\text{LCl}]\text{OH}(\text{H}_2\text{O})_{n-1}\}^+ \xrightarrow{k} \{[\text{Coen}_2\text{LOH}]\text{Cl}(\text{H}_2\text{O})_{n-1}\}^+$$

where the species in the coordination shell are enclosed by square brackets and those in the inner solvation shell by { }. Aquation may be regarded in the same way as a rearrangement between the solvation and coordination shells:

$$\{[\text{Coen}_2\text{LCl}](\text{H}_2\text{O})_n\}^{2+} + \text{H}_2\text{O} \overset{K}{\rightleftharpoons} \{[\text{Coen}_2\text{LCl}]\text{H}_2\text{O}(\text{H}_2\text{O})_{n-1}\}^{2+} + \text{H}_2\text{O}$$

$$\{[\text{Coen}_2\text{LCl}]\text{H}_2\text{O}(\text{H}_2\text{O})_{n-1}\}^{2+} \xrightarrow{k} \{[\text{Coen}_2\text{LH}_2\text{O}]\text{Cl}(\text{H}_2\text{O})_{n-1}\}^{2+}$$

The attack of the hydroxide ion on the inner coordination shell may well proceed by a Grotthus chain-transfer.[23] The solvation shell, consisting of tightly-bound and well-ordered solvent molecules, acts as a physical barrier against attack by an incoming group, but the hydroxide ion, by removal of a proton from a solvent molecule, can generate a new hydroxide ion which in turn repeats the process. Consequently, attack on the metal takes place without physical penetration of the barrier.

Reactions without metal–ligand bond cleavage

Up to this point all the substitution reactions discussed have involved cleavage of the bond between metal and ligand. However, it is known that some replacements occur with preservation of the metal–ligand bond. A familar example in preparative coordination chemistry is the rapid conversion of carbonatoamminecobalt(III) complexes, like $[\text{Co(NH}_3)_5\text{CO}_3]^+$, into the corresponding aquo complexes by the addition of excess acid

$$[(\text{NH}_3)_5\text{Co—O—CO}_2]^+ + 2\text{H}^+ \rightarrow [(\text{NH}_3)_5\text{Co—OH}_2]^{3+} + \text{CO}_2$$

By performing the reaction in the presence of ^{18}O-labelled water it has been shown that it is the O—C bond, rather than the Co—O bond, which breaks, since both products show no uptake of ^{18}O. This result receives confirmation from the observation that reactions involving fission of Co—O bonds are invariably slow. In all probability the reaction proceeds along the following route. Firstly, a series of equilibria occur involving the bicarbonato species $[Co(NH_3)_5CO_3H]^{2+}$

$$(NH_3)_5Co{-}O{-}C\genfrac{}{}{0pt}{}{\overset{..}{=}O^{+}}{\underset{..}{=}O} + H^+ \overset{\text{fast}}{\rightleftharpoons} (NH_3)_5Co{-}O{-}C\genfrac{}{}{0pt}{}{\overset{..}{=}O^{2+}}{\diagdown OH}$$

$$\overset{\text{fast}}{\rightleftharpoons} (NH_3)_5Co{-}\overset{H}{O}{-}C\genfrac{}{}{0pt}{}{\overset{..}{=}O^{2+}}{\underset{..}{=}O}$$

Decarboxylation then occurs as the rate-determining step

$$(NH_3)_5Co{-}\overset{H}{O}{-}C\genfrac{}{}{0pt}{}{\overset{..}{=}O^{2+}}{\underset{..}{=}O} \xrightarrow{\text{slow}} [(NH_3)_5Co{-}OH]^{2+} + CO_2$$

followed by

$$[(NH_3)_5Co{-}OH]^{2+} + H^+ \overset{\text{fast}}{\rightleftharpoons} [(NH_3)_5Co{-}OH_2]^{3+}$$

The reaction

$$[Co(NH_3)_5Cl]^{2+} + NO_2^- \rightleftharpoons [Co(NH_3)_5NO_2]^{2+} + Cl^-$$

has been studied in considerable detail. The formation of the nitro complex occurs indirectly through the sequence: chloro complex → aquo complex → nitrito complex → nitro complex. Isotopic tracer experiments have established that the Co—O bond of the aquo complex remains intact in the formation of the nitrito form since the reaction

$$[(NH_3)_5Co{-}^{18}OH_2]^{3+} + NO_2^- \xrightarrow{H_2O} [(NH_3)_5Co{-}^{18}ONO]^{2+} + H_2O$$

takes place with complete retention of ^{18}O. Kinetic studies[24] in nitrite–nitrous acid buffers at pH 4–5 yield the empirical rate law

$$\text{rate} = d[\textit{nitrito complex}]/dt = k[\textit{aquo complex}][NO_2^-]^2[H^+]$$

which is equivalent kinetically to

$$\text{rate} = k'[\textit{hydroxo complex}][NO_2^-]^2[H^+]^2$$

as a consequence of the aquo complex existing in acid-base equilibrium with the hydroxo form. This type of rate expression is found for the nitrosation of amines and has been interpreted there as indicating nitrosyl ion attack on the amido nitrogen. It is possible that a similar attack occurs on the coordinated hydroxy group of the hydroxo complex. The mechanistic scheme for the overall process is accordingly

$$[(NH_3)_5Co{-}Cl]^{2+} + H_2O \rightarrow [(NH_3)_5Co{-}OH_2]^{3+} + Cl^-$$

$$[(NH_3)_5Co{-}OH_2]^{3+} \rightleftharpoons [(NH_3)_5Co{-}OH]^{2+} + H^+$$

$$2HNO_2 \rightleftharpoons N_2O_3 + H_2O$$

$$[(NH_3)_5Co{-}OH]^{2+} + N_2O_3 \rightleftharpoons (NH_3)_5Co{-}O{\cdots}H^+ \quad (O{-}N{\cdots}NO_2^- \text{ bonded to O})$$

$$\rightarrow [(NH_3)_5Co{-}ONO]^{2+} + HNO_2 \quad \text{rate-determining}$$

$$[(NH_3)_5Co{-}ONO]^{2+} \rightleftharpoons [(NH_3)_5Co{-}NO_2]^{2+}$$

It is interesting that recognition of the above mechanism has enabled nitrito–nitro linkage isomers of Rh(III), Ir(III), and Pt(IV) to be prepared, which previously had eluded synthesis.

Reactions of platinum(IV) complexes

On electrostatic grounds (see Table 2.1), increasing the charge and size of the central metal atom in a complex would be expected to favour reaction via an associative path. Since Co(III) complexes react, in general, by a dissociative process, there is a likelihood that a comparison of the substitution reactions of the d^6 systems, Co(III) and Pt(IV), might reveal significant differences in mechanism. Unfortunately, due to the strength of the metal–ligand bond, octahedral complexes of Pt(IV) are inert to substitution unless catalysed by Pt(II).

The direct exchange of chloride (as ^{36}Cl) with *trans*-$[Pt(en)_2Cl_2]^{2+}$ is extremely slow and is almost completely inhibited by the removal of trace quantities of Pt(II), present as impurities in the Pt(IV) complex (prepared by chlorination of $[Pt(en)_2]^{2+}$). In the presence of added amounts of Pt(II) the rate law takes the form[25]

$$\text{rate} = k[Pt(en)_2Cl_2{}^{2+}][Pt(en)_2{}^{2+}][Cl^-]$$

The two-stage mechanism proposed to explain these results is

$$[Pt(en)_2]^{2+} + Cl^- \overset{\text{fast}}{\rightleftharpoons} [Pt(en)_2Cl]^+$$

$$\begin{array}{c}\text{en}\\ \text{Cl—Pt—Cl}^{2+}\\ \text{en}\end{array} + \begin{array}{c}\text{en}\\ \text{Pt—Cl}^{+}\\ \text{en}\end{array} \overset{\text{slow}}{\rightleftharpoons} \begin{array}{c}\text{en}\qquad\text{en}\\ \text{Cl—Pt}\cdots\text{Cl}\cdots\text{Pt—Cl}^{3+}\\ \text{en}\qquad\text{en}\end{array} \overset{\text{slow}}{\rightleftharpoons}$$

$$\begin{array}{c}\text{en}\\ \text{Cl—Pt}^{+}\\ \text{en}\end{array} + \begin{array}{c}\text{en}\\ \text{Cl—Pt—Cl}^{2+}\\ \text{en}\end{array}$$

As regards the first step, there is considerable evidence to suggest that penta-coordinated species of square-planar complexes are feasible. No spectrophotometric evidence could be found for the bridged intermediate, through which chloride exchange takes place. However, the existence of such a complex in the solid state is substantiated by the observation that evaporation of a colourless aqueous solution containing equivalent amounts of *trans*-$[Pt(en)_2Cl_2]Cl_2$ and $[Pt(en)_2]Cl_2$ results in the formation of an orange solid which presumably has the empirical formula $[Pt_2(en)_4Cl_3]Cl_3$ with a structure given by

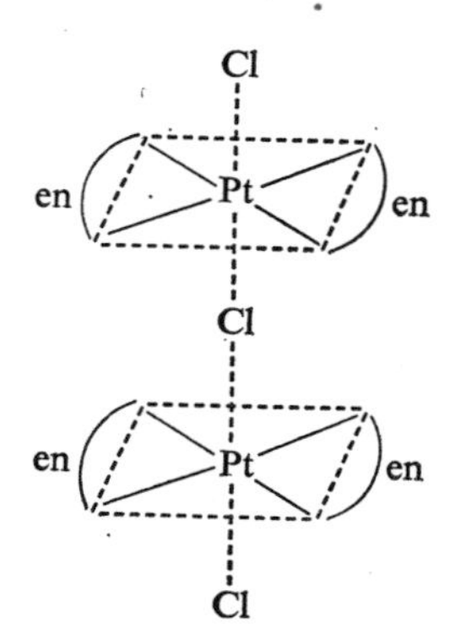

As required by the mechanism the rate of platinum exchange (using ^{195}Pt as a tracer) is the same as that for chloride ion exchange. It is significant that the only chloroammine platinum(IV) complex investigated that does not exchange chloride ion in the presence of Pt(II) is the tetramethylethylenediamine complex, *trans*-$[Pt(tetra\text{-}mean)_2Cl_2]^{2+}$. In this case a bridged complex cannot be formed through the chloro group because of the bulkiness of the C-methyl groups on the chelate ring.

The direct reaction

$$trans\text{-}[Pt(en)_2Cl_2]^{2+} + NO_2^- \rightarrow trans\text{-}[Pt(en)_2NO_2Cl]^{2+} + Cl^-$$

takes place only slowly and shows a marked period of induction. The postulated mechanism[26] has as the first step the reduction of Pt(IV) to Pt(II) by nitrite ions

$$[Pt(en)_2Cl_2]^{2+} + NO_2^- + H_2O \rightarrow [Pt(en)_2]^{2+} + 2H^+ + 2Cl^- + NO_3^-$$

followed by

$$[Pt(en)_2]^{2+} + NO_2^- \underset{}{\overset{\text{fast}}{\rightleftharpoons}} [Pt(en)_2NO_2]^+$$

$$\underset{\text{en}}{\overset{\text{en}}{Cl\text{—}Pt\text{—}Cl^{2+}}} + \underset{\text{en}}{\overset{\text{en}}{Pt\text{—}NO_2^+}} \xrightarrow{\text{slow}} \underset{\text{en}\quad\quad\text{en}}{\overset{\text{en}\quad\quad\text{en}}{Cl\text{—}Pt\cdots Cl\cdots Pt\text{—}NO_2^{3+}}} \xrightarrow{\text{fast}}$$

$$Cl^- + [Pt(en)_2]^{2+} + \underset{\text{en}}{\overset{\text{en}}{Cl\text{—}Pt\text{—}NO_2^{2+}}}$$

With the addition of catalytic amounts of Pt(II), as $[Pt(en)_2]^{2+}$, the induction period disappears and the reaction is greatly accelerated.

Reactions in non-aqueous solvents

The first systematic study on octahedral substitution in non-aqueous media[27] was concerned with the reaction between *cis*-$[Co(en)_2Cl_2]^+$ and a number of anions in methanol solvent. It is likely that such reactions proceed without the intervention of a solvento intermediate, $[Co(en)_2(CH_3OH)Cl]^{2+}$, since this is too labile to have an independent existence. The anions studied fall into two groups: (*a*) those which enter at a rate dependent on their concentration, e.g., CH_3O^-, N_3^-, NO_2^-, and CH_3COO^-; (*b*) less reactive reagents which enter at a common rate which is independent of concentration, e.g., SCN^-, Br^-, Cl^-, and NO_3^-. A duality of mechanism was invoked to account for these differences in behaviour. Anions in group (*b*) were assumed to react via a dissociative process involving a common intermediate, $[Co(en)_2Cl]^{2+}$, whose rate of formation determined the rate of the overall substitution. Those ligands in group (*a*), considered to be better nucleophiles, reacted via an associative process. However, further investigations have shown that the increased reactivity and concentration dependence of N_3^-, NO_2^-, and CH_3COO^- is due to reaction of these basic ligands with the solvent (*solvolysis*) to produce methoxide ions

$$CH_3OH + L \rightleftharpoons CH_3O^- + HL$$

The methoxide ion, analogous to the hydroxide ion in water, then reacts via a S_N1CB mechanism.

Complications due to solvolysis can be avoided by the choice of a non-protonic solvent, such as dimethylsulphoxide. A good illustration occurs in the reactions of the *trans* isomers of chloropentamine-cobalt (III) complexes with nitrite, azide or thiocyanate,[28] for example

$$[Co(en)_2ACl]^+ + NO_2^- \xrightarrow[DMS]{} [Co(en)_2ANO_2]^+ + Cl^- \qquad \text{slow}$$

Such reactions are very slow but are catalysed markedly by small quantities of hydroxide ion which, however, do not affect the nature of the product

$$[Co(en)_2ACl]^+ + NO_2^- \xrightarrow[DMS]{OH^-} [Co(en)_2ANO_2]^+ + Cl^- \qquad \text{very fast}$$

The function of the OH^- ion could involve a direct displacement

$$[Co(en)_2ACl]^+ + OH^- \rightarrow [Co(en)_2AOH]^+ + Cl^-$$

followed by

$$[Co(en)_2AOH]^+ + NO_2^- \rightarrow [Co(en)_2ANO_2]^+ + OH^- \qquad \text{very slow}$$

However, this possibility is ruled out by the observation that the latter reaction is too slow to account for the catalytic effect of hydroxide ion. The alternative is a S_N1CB scheme, in which a five-coordinated intermediate is formed*

$$[Co(en)_2ACl]^+ + OH^- \underset{}{\overset{fast}{\rightleftharpoons}} [Co(en)(en - H)ACl] + H_2O$$

$$[Co(en)(en - H)ACl] \xrightarrow{slow} [Co(en)(en - H)A]^+ + Cl^-$$

which then reacts rapidly with the incoming ligand

$$[Co(en)(en - H)A]^+ + NO_2^- \xrightarrow{fast} [Co(en)(en - H)ANO_2]$$

* en – H represents ethylenediamine *minus* a hydrogen atom.

A rapid proton-transfer gives the final product and regenerates hydroxide ion

$$[Co(en)(en - H)ANO_2] + H_2O \xrightarrow{\text{fast}} [Co(en)_2ANO_2]^+ + OH^-$$

On this mechanism the rate of the overall reaction is governed by the rate of formation of the intermediate, $[Co(en)(en - H)A]^+$, and hence should depend only on the concentration of the base and be

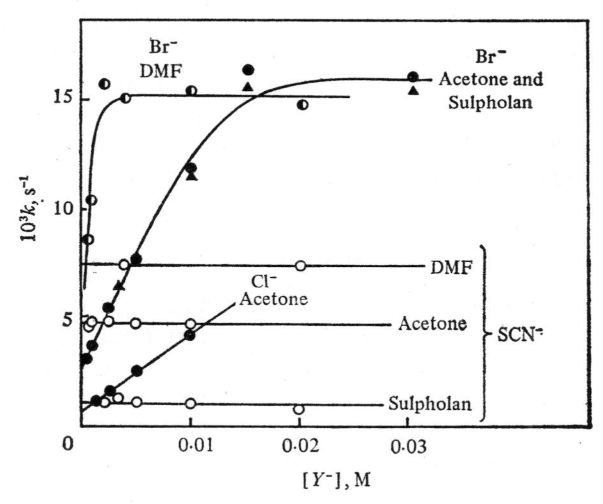

Fig. 2.4. Rates of replacement of water in *trans*-$[Co(en)_2NO_2H_2O](ClO_4)_2$ by Cl^-, Br^-, and SCN^- in non-aqueous solvents at 25° as a function of anion concentration. From M. L. Tobe, in *Mechanisms of Inorganic Reactions,* Advances in Chemistry Series (ed. R. F. Gould), No. 49, p. 12, American Chemical Society, 1965.

independent of the nature and concentration of the incoming ligand. In fact, nitrite reacted at the same rate as thiocyanate and azide under the same conditions of base (hydroxide ion or piperidine).

A recent investigation[29] of the anation reactions of *trans*-$[Co(en)_2NO_2H_2O]^{2+}$ in a variety of solvents (acetone, sulpholan* and dimethylformamide) has given rise to some interesting fine points of interpretation. The kinetic results are summarized by Fig. 2.4. Conductometric measurements revealed that ion aggregates exist under these conditions: thiocyanate over the whole concentration range forms an ion triplet of the type $\{[Co(en)_2NO_2H_2O]^{2+}\text{---}2SCN^-\}$

* Tetrahydrothiophen-1,1-dioxide.

whereas chloride and bromide form similar 1:2 aggregates at low concentrations but have the ability to add on to form 1:3 aggregates at higher concentrations. Evidently Fig. 2.4 can be explained on the basis of the differing reactivities of the various aggregates. Thiocyanate reacts at a rate independent of its concentration in all three solvents. Chloride and bromide show an anion-dependence at low concentrations leading, at sufficiently high concentrations, to a limiting rate.

The motive for carrying out kinetic studies of substitution reactions in non-aqueous solvents was, originally, the hope that the information gleaned might be of assistance in elucidating the mechanism of substitution in aqueous media. Experience has shown, however, that each solvent system must be considered unique. Also, although there is currently a great deal of activity in this field, it is true to say that the interpretation of the results is beset with conceptual difficulties.

Fast reactions[30]

The replacement of water from hydrated metal ions by an anionic ligand has been studied by a number of fast reaction techniques, notably ultrasonic absorption. The kinetic data have been interpreted in terms of a scheme in which an outer-sphere complex intervenes prior to the production of an inner-sphere complex. As noted already, such a mechanism is prevalent for oppositely-charged reactants

$$[M(H_2O)_n]^{m+} + Y^- \rightleftharpoons [M(H_2O)_n]^{m+}.Y^-$$

$$[M(H_2O)_n]^{m+}.Y^- \rightleftharpoons [M(H_2O)_{n-1}Y]^{(m-1)+} + H_2O$$

The rate of formation of the outer-sphere complex, being determined largely by electrostatic factors, is, to a great extent, independent of the electronic configuration of the metal ion and of the nature of the ligand. Two distinct processes are often realized in the formation of the outer-sphere complex

$$[M(H_2O)_n]^{m+} + Y^- \underset{k_a'}{\overset{k_a}{\rightleftharpoons}} [M(H_2O)_n]^{m+}.H_2O.Y^-$$

$$[M(H_2O)_n]^{m+}.H_2O.Y^- \underset{k_b'}{\overset{k_b}{\rightleftharpoons}} [M(H_2O)_n]^{m+}.Y^-$$

The preliminary stage is the diffusion of the reactants together to form a loose complex in which the metal ion and entering reagent are separated by more than one water molecule. Such a process is extremely rapid, with $k_a \sim 10^9$–10^{10} M^{-1} s^{-1} and $k_a' \sim 10^8$–10^{10} s^{-1}. The second stage is the subsequent formation of a complex in which metal ion and reagent are separated only by a single water molecule ($k_b \sim 10^8$–10^9 s^{-1} and $k_b' \sim 10^7$–10^{10} s^{-1}).

The much slower rate of formation of an inner-sphere complex is determined primarily by the electronic configuration of the metal ion,

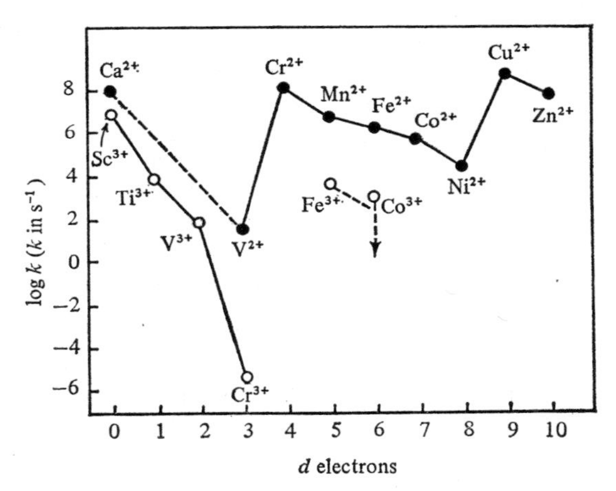

Fig. 2.5. Plot of log of rate constants (replacement of a coordinated water molecule by a ligand) versus number of *d*-electrons for bi- and tri-valent transition metal ions (substitution by SO_4^{2-}, except for V^{2+} and Cr^{2+} where the ligands are SCN^- and bipy, respectively; values for Cr^{3+}, Fe^{3+}, and Co^{3+} are for water exchange). From data kindly supplied by M. Eigen and H. Diebler.

and is, in general, insensitive to the nature of the entering group. Thus the rate constants for the reaction of Ni^{2+} with a wide variety of ligands (bidentate and monodentate) are remarkably similar and lie around 10^4 M^{-1} s^{-1} at 25°. Positively-charged ligands react more slowly and ligands with high negative charges react more quickly. Significantly the rate constant for the water exchange of Ni^{2+} has a value of 3×10^4 M^{-1} s^{-1}. Indeed, for bivalent metal ions, there is a close resemblance between the rates of water exchange and complex formation. This observation suggests that the rate-controlling step in both types of processes is the replacement of a water molecule in the inner sphere by a species from the outer sphere.

A comparison (see Fig. 2.5) of the rates of replacement of a water molecule by a ligand (i.e., transformation of an outer-sphere complex into an inner-sphere complex) reveals the following order of reactivity for bivalent metal ions:

$$V^{2+} < Ni^{2+} < Co^{2+} < Fe^{2+} < Mn^{2+} < Zn^{2+} < Cr^{2+} \sim Ca^{2+} < Cu^{2+}$$

i.e.,

$$d^3 < d^8 < d^7 < d^6 < d^5 < d^{10} < d^4 \sim d^0 < d^9$$

A confirmatory, but more limited, series is obtained on comparing the rates of water exchange from n.m.r. data:

$$Ni^{2+} < Co^{2+} < Fe^{2+} < Mn^{2+} < Cu^{2+}$$

The following points are worth particular attention.

(*a*) The relatively slow rates of reaction of the d^3 system of V^{2+} and the d^8 system of Ni^{2+} are in line with the predictions of crystal field theory (see p. 23).

(*b*) The extremely rapid rates of Cr^{2+} (d^4) and Cu^{2+} (d^9) are attributed to Jahn–Teller effects. Because of the distortion of the octahedral structures, the water molecules occupying the axial positions are held less firmly than those occupying equatorial positions and therefore exchange more readily with the solvent. Anomalously slow reactions (found with multidentate ligands) probably result from the substitution of equatorial water molecules. In general, it is thought that configurational changes are necessary before the equatorial molecules can be replaced. To accomplish this, an inversion takes place in which two equatorial positions become axial and two axial positions become equatorial. The axial positions, labilized by Jahn–Teller distortions, then exchange rapidly.

(*c*) Rapid rates are noted also for the d^{10} systems of Hg^{2+}, Cd^{2+}, and Zn^{2+}. These metal ions, along with cobalt(II), form tetrahedral as well as octahedral complexes and at one time it was thought possible that the formation of octahedral complexes occurred through a more active tetrahedral intermediate. However, this is unlikely since, for Zn(II) and Co(II), changes in coordination are too slow to allow such catalytic effects.

Although somewhat complicated by hydrolysis effects, the rates of formation of complexes of trivalent metal ions are determined by the rate of loss of a coordinated water molecule. The order of reactivity for the hexa-aquo ions is given by

$$Rh^{3+} < Cr^{3+} < V^{3+} < Co^{3+} < Fe^{3+} < Ti^{3+}$$

(cf. Fig. 2.5). Studies of the aquopentachloro complexes show that Ir(III) < Rh(III) < Ru(III). The relatively slow substitutions of water in the d^3 ion, $[Cr(H_2O)_6]^{3+}$ and low-spin d^6 ions, $[Rh(H_2O)_6]^{3+}$, $[Co(H_2O)_6]^{3+}$, $[IrCl_5H_2O]^{2-}$, and $[RhCl_5H_2O]^{2-}$ are to be expected in terms of crystal field theory.

A particularly revealing study is that by Sutin[31] on substitution in Co(III). Figure 2.6 is a photograph of an oscilloscope trace showing

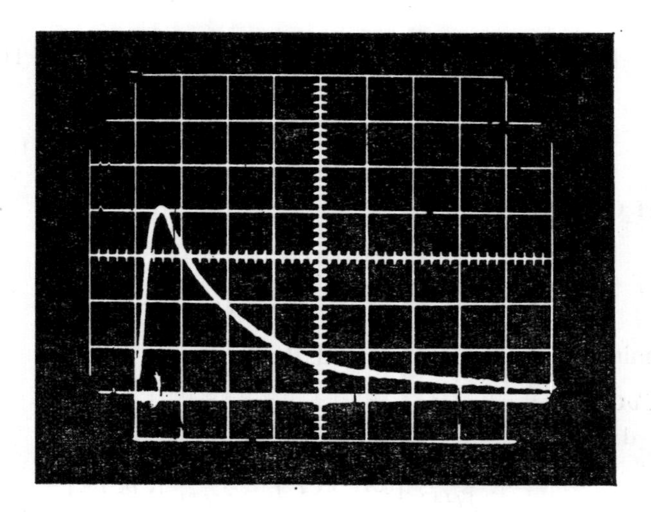

Fig. 2.6. Oscilloscope trace of absorbance (at 280 mμ) of Co(III)-chloride system as a function of time; $[Co(III)] = 7{\cdot}5 \times 10^{-5}$ M, $[Co(II)] = 2{\cdot}4 \times 10^{-4}$ M, $[Cl^-] = 5{\cdot}0 \times 10^{-2}$ M, $[HClO_4] = 1{\cdot}0$ M, 25°. The abscissa scale is 500 ms per major division and the ordinate scale is in arbitrary units of absorbance. (The lower trace corresponds to the absorbance of the solution at 'infinite' time.) From T. J. Conocchioli, G. H. Nancollas, and N. Sutin, *Inorg. Chem.*, 1966, **5**, 1.

the changes taking place at 280 mμ on mixing solutions of Co(III) and chloride ions (each maintained at the same acidity and ionic strength). The initial rapid increase in absorbance and the subsequent slow decrease are due to the formation of the monochloro complex of Co(III) and the reduction of Co(III) by chloride ion, respectively. The latter reaction can be prevented, to a great extent, by using low chloride concentrations at high acidities and high Co(II) concentrations. Under these conditions the formation of $CoCl^{2+}$ is complete before reduction of Co(III) occurs. At the higher acidities the observed dependence of the rate on Cl^-, H^+ and Co(II) concentrations is in accord with a mechanism involving the following equilibria:

$$[Co(H_2O)_6]^{3+} + Cl^- \underset{k_{1d}}{\overset{k_{1f}}{\rightleftharpoons}} [(H_2O)_5CoCl]^{2+} + H_2O \quad K_1$$

$$[Co(H_2O)_6]^{3+} \rightleftharpoons [(H_2O)_5CoOH]^{2+} + H^+ \quad K_{1h}$$

$$[(H_2O)_5CoOH]^{2+} + Cl^- \underset{k_{2d}}{\overset{k_{2f}}{\rightleftharpoons}} [(H_2O)_4CoOHCl]^+ + H_2O \quad K_2$$

$$[Co(H_2O)_5Cl]^{2+} \rightleftharpoons [(H_2O)_4CoOHCl]^+ + H^+ \quad K_{2h}$$

$$[Co(H_2O)_6]^{3+} + [(H_2O)_5CoCl]^+ \underset{k_{3d}}{\overset{k_{3f}}{\rightleftharpoons}} [Co(H_2O)_6]^{2+} + [(H_2O)_5CoCl]^{2+} \quad K_3$$

$$[Co(H_2O)_6]^{2+} + Cl^- \rightleftharpoons [(H_2O)_5CoCl]^+ + H_2O \quad K_4$$

Assuming the reactions to be rapid, then

$$\frac{d[CoCl^{2+}]}{dt} = k_{1f}[Co^{3+}][Cl^-] - k_{1d}[CoCl^{2+}] + \frac{k_{2f}K_{1h}}{[H^+]}[Co^{3+}][Cl^-] - \frac{k_{2d}K_{2h}}{[H^+]}[CoCl^{2+}] + k_{3f}K_4[Co^{2+}][Co^{3+}][Cl^-] - k_{3d}[Co^{2+}][CoCl^{2+}] \quad (2.1)$$

At equilibrium $d[CoCl^{2+}]/dt = 0$, and it follows that

$$k_{1f} = \left(k_{1d} + \frac{k_{2d}K_{2h}}{[H^+]} + k_{3d}[Co^{2+}]\right)\frac{[CoCl^{2+}]_{eq}}{[Co^{3+}]_{eq}[Cl^-]} - \left(k_{3f}K_4[Co^{2+}] + \frac{k_{2f}K_{1h}}{[H^+]}\right) \quad (2.2)$$

where $[H^+]$, $[Cl^-]$, and $[Co^{2+}]$ are now equilibrium concentrations. Also, by mass balance

$$[Co^{3+}] + [CoCl^{2+}] + [CoOH^{2+}] = [Co^{3+}]_{eq} + [CoCl^{2+}]_{eq} + [CoOH^{2+}]_{eq} \quad (2.3)$$

if $CoOHCl^{+}$ and other hydrolysed species of cobalt(III) are neglected. By substituting k_{1f} and $[Co^{3+}]$, (2.1) becomes

$$\frac{d[CoCl^{2+}]}{dt} = \left\{\left(\frac{k_{1f}[H^+] + k_{2f}K_{1h} + k_{3f}K_4[Co^{2+}][H^+]}{[H^+] + K_{1h}}\right)[Cl^-] + \left(\frac{k_{1d}[H^+] + k_{2d}K_{2h} + k_{3d}[Co^{2+}][H^+]}{[H^+]}\right)\right\} \times [CoCl^{2+}]_{eq} - [CoCl^{2+}] \quad (2.4)$$

Thus the observed first-order rate constant is given by

$$k_{obs} = k'[Cl^-] + k'' \quad (2.5)$$

where

$$k' = \frac{k_{1f}[H^+] + k_{2f}K_{1h} + k_{3f}K_4[Co^{2+}][H^+]}{[H^+] + K_{1h}}$$
$$k'' = \frac{k_{1d}[H^+] + k_{2d}K_{2h} + k_{3d}[Co^{2+}][H^+]}{[H^+]} \quad (2.6)$$

Values for k' and k'' are obtained from the slopes and intercepts of linear plots of k_{obs} versus $[Cl^-]$, for various acidities at constant Co(II) concentration. Furthermore, (2.6) can be rearranged in the form

$$k''/k' = K_{1h}/K_1[H^+] + 1/K_1 \quad (2.7)$$

allowing K_{1h} and K_1 to be evaluated from the slope and intercept of a plot of k''/k' versus $1/[H^+]$. Again from (2.6) plots of k'' versus $1/[H^+]$ and $k'(1 + K_{1h}/[H^+])$ versus $1/[H^+]$ yield values for $k_{2f}K_{1h}$ (and therefore k_{2f}) and $k_{2d}K_{2h}$, respectively. Finally from the observed rate law (2.5) written in the form

$$k_{obs} = \left(\frac{k_{1f}[H^+] + k_{2f}K_{1h}}{[H^+] + K_{1h}}\right)[Cl^-] + \left(\frac{k_{1d}[H^+] + k_{2d}K_{2h} + k_{3d}[H^+]}{[H^+]}\right) + \left(\frac{k_{3f}K_4[H^+][Cl^-]}{[H^+] + K_{1h}} + k_{3d}\right)[Co^{2+}] \quad (2.8)$$

and the corresponding plot of k_{obs} versus $[Co^{2+}]$, at constant $[H^+]$ and $[Cl^-]$, values of k_{3d} and $k_{3f}K_4$ are determined which on substitution into (2.6) give limits for k_{1f} and k_{1d}. The evaluated constants are collected in Table 2.12 along with comparable data for the Fe(III)-chloride system (cf. Table 2.13).

The sound absorption method is the only means whereby the very high substitution rates of alkali metal ions can be obtained. Since the usual inorganic-type ligands form very weak complexes, strongly-complexing reagents, like aminopolycarboxylic acids (EDTA and iminodiacetate), nitrilotriacetate, and adenosine triphosphate, had

Table 2.12

Comparison of the cobalt(III)-chloride and iron(III)-chloride systems at 25·0°; ionic strength 3·0 M

	Co(III) system*	Fe(III) system†
K_1, M^{-1}	26 ± 3	8·2
k_{1f}, M^{-1} s^{-1}	$\leqslant 2$	9 ± 2
k_{1d}, s^{-1}	$\leqslant 0{\cdot}05$	$1{\cdot}1 \pm 0{\cdot}2$
k_{2f}, M^{-1} s^{-1}	$(2 \pm 1) \times 10^2$	$(2{\cdot}1 \pm 0{\cdot}3) \times 10^4$
$k_{2d}K_{2h}$, M s^{-1}	$1{\cdot}5 \pm 0{\cdot}5$	$3{\cdot}4 \pm 0{\cdot}6$
$k_{2f}K_{1h}$, s^{-1}	$40 \pm 1{\cdot}5$	28 ± 5
k_{3d}, M^{-1} s^{-1}	$1{\cdot}0 \pm 0{\cdot}2$	$12{\cdot}1 \pm 1{\cdot}6$
$k_{3f}K_4$, M^{-2} s^{-1}	26 ± 5	96 ± 12

* From ref. (31).

† From R. J. Campion, T. J. Conocchioli, and N. Sutin, *J. Amer. Chem. Soc.*, 1964, **86**, 4591.

to be employed. The rate-controlling step is the substitution of several water molecules by the attacking ligand. The rates parallel the ionic radii, i.e., $Cs^+ > Rb^+ > K^+ > Na^+ > Li^+$, although the differences are not great. Furthermore, the rate of substitution varies to some extent with the nature of the ligand, the rates being highest for phosphates.

The water-replacement processes for alkaline-earth ions are very rapid also. As expected from considerations of ionic size, the rate order is $Ba^{2+} > Sr^{2+} > Ca^{2+} \gg Mg^{2+}$; the rates for Ba^{2+}, Sr^{2+}, and Ca^{2+} are of the same order as those of the alkali metals. The large difference in reactivity between Mg^{2+} and Ca^{2+} (three orders of magnitude) is striking.

By way of summary, sufficient data are now available to allow comparison to be made of water substitution in the inner coordination shell of metal ions. Broadly speaking, three different mechanisms can be recognized.

(*a*) The rate-determining step is the attack of ligand with formation of a chelate; the alkali and some alkaline-earth metal ions fall into this category. This class is characterized by very high rate constants ($> 10^7\ s^{-1}$) and a dependence on the type of ligand. Zn^{2+}, Cd^{2+}, and Hg^{2+} ions behave anomalously in that they react as rapidly as Ca^{2+} but show a lack of ligand specificity. This difference is most probably due to the fact that they have a closed *d*-shell (d^{10}).

(*b*) The rate-determining step is the loss of coordinated water as typified by Mg^{2+} and most divalent metal ions of the first transition series. In this category the rate constants are less than $10^7\ s^{-1}$, and the rates show little or no dependence on the type of ligand, being characteristic of the metal ion alone.

(*c*) The rate-determining step is the hydrolysis of a water molecule in the coordination shell

$$[M\ H_2O)^{n+} + B^- \rightarrow (M\ OH)^{(n-1)+} + BH$$

The rates are low and are influenced by the type of ligand: a correlation between rate and basicity of reagent is to be expected. Fe^{3+}, Al^{3+}, and Be^{2+} are members of this class. The kinetics of the formation of complexes between Fe^{3+} and various simple anions have received detailed study by a number of fast reaction techniques (pressure-jump, baffle flow, and continuous flow). The rate law found for the formation of the intense red thiocyanate complex is

$$\mathrm{d}[FeNCS^{2+}]/\mathrm{d}t = k_1[Fe^{3+}][SCN^-] + k_2[Fe^{3+}][SCN^-]/[H^+]$$

The existence of the second term implies that, as well as the direct reaction with Fe^{3+}, a reaction between $FeOH^{2+}$ and SCN^- takes place. This form of rate law is common to the other systems investigated. The suggestion has been made that the initial step in these reactions is the loss of a proton from the hydrated Fe(III) ion, for example

$$[Fe(H_2O)_5OH_2]^{3+} + F^- \rightarrow [Fe(H_2O)_5OH]^{2+} + HF$$

The rate data given in Table 2.13 show that, up to a point, the rate can be related to the basicity of the anion which increases in the order $Br^- < Cl^- < SCN^- < SO_4^{2-} < F^-$.

Table 2.13

Rate constants at 25° for reaction of Fe(III) with various anions in aqueous solution

Anion (Y)	k_1, $M^{-1}\ s^{-1}$ ($Fe^{3+} + Y$)	k_2, s^{-1} ($FeOH^{2+} + Y$)
Br^-*	20 ± 6 (22°)	31 ± 8 (22°)
Cl^-†	$9{\cdot}4 \pm 1$	18 ± 2
SCN^-†	127 ± 10	20 ± 2
SO_4^{2-}	$6{\cdot}4 \times 10^3$‡	720 ± 40§
F^-	1×10^4¶	

* From P. Matthies and H. Wendt, *Z. phys. Chem. (Frankfurt)*, 1961, **30**, 137.

† From J. F. Below, R. E. Connick, and C. P. Coppel, *J. Amer. Chem. Soc.*, 1958, **80**, 2961; R. E. Connick and C. P. Coppel, *J. Amer. Chem. Soc.*, 1959, **81**, 6389.

‡ From G. G. Davis and W. MacF. Smith, *Canad. J. Chem.*, 1962, **40**, 1836.

§ From H. Wendt and H. Strehlow, *Z. Elektrochem.*, 1962, **66**, 228.

¶ From D. Pouli and W. MacF. Smith, *Canad. J. Chem.*, 1960, **38**, 567.

B. Square-planar complexes

Square-planar complexes are formed by the d^8 systems of Pt(II), Pd(II), Ni(II), Au(III), Rh(I), and Ir(I). The kinetics of ligand substitution reactions of the inert and stable complexes of platinum(II) have been examined in greatest detail. For reactions such as

$$[Pt(NH_3)_3Cl]^+ + Y^- \rightarrow [Pt(NH_3)_3Y]^+ + Cl^-$$

the kinetics are described by a two-term rate law which has the form

$$\text{rate} = k_S[complex] + k_Y[complex][Y^-]$$

The composite nature of the rate law implies that such reactions proceed via two parallel routes: a solvent path

$$[Pt(NH_3)_3Cl]^+ + H_2O \underset{}{\overset{\text{slow},\ k_S}{\rightleftharpoons}} [Pt(NH_3)_3H_2O]^{2+} + Cl^-$$

$$[Pt(NH_3)_3H_2O]^{2+} + Y^- \xrightarrow{\text{fast}} [Pt(NH_3)_3Y]^+ + H_2O$$

and a direct displacement of ligand by the incoming nucleophile

$$[Pt(NH_3)_3Cl]^+ + Y^- \underset{}{\overset{\text{slow}, k_Y}{\rightleftharpoons}} \left[(NH_3)_3Pt\begin{smallmatrix} Y \\ Cl \end{smallmatrix}\right] \rightarrow [Pt(NH_3)_3Y]^+ + Cl^-$$

There are valid reasons for believing that both routes involve associative processes. It will suffice at this stage to cite two sorts of evidence. Firstly, variation of the charge of the complex has only a slight effect on the rate. This is best illustrated by the rate data for aquation and chloride exchange of a series of chloroammine complexes, the charge on the substrate varying from –2 to +1 (see Table 2.14).

Table 2.14

Rate data* (at 25° and $\mu = 0{\cdot}32$ M) for aquation† and chloride exchange of chloroammineplatinum(II) complexes

Complex	$10^5 k_{H_2O}$, s^{-1}	$10^5 k_{Cl^-}$, $M^{-1}\ s^{-1}$
$[PtCl_4]^{2-}$	3·9	<3
$[Pt(NH_3)Cl_3]^-$§	5·6	<3
cis-$[Pt(NH_3)_2Cl_2]$	2·5	~3
trans-$[Pt(NH_3)_2Cl_2]$	9·8	78
$[Pt(NH_3)_3Cl]^+$	2·6	7

* From M. A. Tucker, C. B. Colvin, and D. S. Martin, *Inorg. Chem.*, 1964, 3, 1373.
† Replacement of first chloride ion.
§ Replacement of *cis* Cl.

These results are in striking contrast to the behaviour of octahedral complexes where charge has a drastic effect on reactivity: for example, *trans*-$[Co(NH_3)_4Cl_2]^+$ hydrolyses approximately 1000 times faster than $[Co(NH_3)_5Cl]^{2+}$. Insensitivity of rate to the charge on the complex is a characteristic of associative processes where bond making and bond breaking play roles of comparable importance (see Table 2.1, p. 21). Secondly, an increase in steric hindrance in the complex is accompanied by a decrease in reactivity. For example, the series of *cis*- and *trans*-$[Pt(PEt_3)_2RCl]$ complexes show, in their reaction with pyridine in ethanol solution, a marked decrease in rate as the group R is varied from phenyl to *o*-tolyl to mesityl: relative rates for the *cis*- and *trans*-isomers are, respectively, 100,000:200:1 and 30:6:1.[32] This effect is to be expected if reaction necessitates an increase in coordination number. In contrast, it will be recalled (p. 26) that the dissociative reactions of octahedral Co(III) complexes exhibit a steric acceleration. In general, as we shall see in later sections, substitution reactions of Pt(II) complexes display other features consistent with an associative mechanism.

Nucleophilic reactivity

The rates of substitution reactions of six-coordinated metal complexes are, in general, independent of the nature of the substituent (known exceptions are discussed on pp. 38–40). However, kinetic information has given direct indication of the factors governing reagent reactivities for the case of substitutions in square-planar complexes where the rate is reagent-dependent. The substitution reactions of *trans*-$[Pt(py)_2Cl_2]$ with various nucleophiles (Y) in methanol solution have received a detailed study.[33] The rates of reaction conform to the expression

$$\text{rate} = k_S[Pt(py)_2Cl_2] + k_Y[Pt(py)_2Cl_2][Y]$$

that is

$$\text{rate} = (k_S + k_Y[Y])[Pt(py)_2Cl_2]$$

and, under conditions of excess Y concentration, the observed first-order rate constant is defined as $k_{obs} = k_S + k_Y[Y]$. A plot of k_{obs} versus $[Y]$ is given in Fig. 2.7 for Y = SCN^-, I^-, C_6H_5SH, Br^-, NH_2OH, N_3^-, and NO_2^-. It is seen that a common intercept (equal to k_S) is attained, the slopes of the lines yielding the k_Y value for the particular substituent. Similar plots have been reported for the reactions (in aqueous solution) of $[Pt(dien)Cl]^+$ (where dien represents diethylenetriamine, $NH(CH_2CH_2NH_2)_2$) with I^-, Br^-, Cl^-, and OH^-, and of $[Pt(dien)Br]^+$ with $SC(NH_2)_2$, SCN^-, I^-, N_3^-, NO_2^-, pyridine, Cl^-, and OH^-.

Since substitution reactions are, in essence, acid–base reactions (in the Lewis sense)

$$\text{M—X (acid-base complex)} + \text{Y (base)} \rightarrow \text{M—Y (acid-base complex)} + \text{X (base)}$$

it would be expected that the effectiveness of the nucleophile (Y) in displacing X (the replaced ligand) from the complex could be measured, to some extent, in terms of the basicity of Y. However, inspection of data for the k_Y and pK_A values of the entering group reveals a total lack of correlation between the rate of attack and the basicity of the nucleophile. Evidently other variables are responsible in deciding the reactivity of the reagent.

The current trend is to relate reactivity with the microscopic polarizability or degree of 'softness' of the nucleophile.[34] A 'soft'

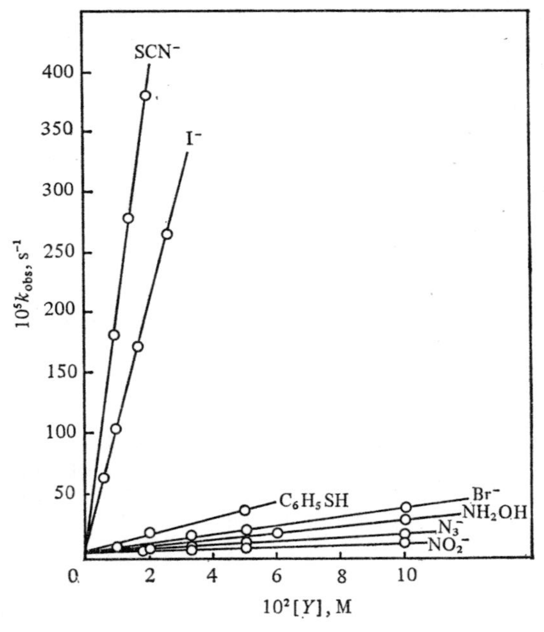

Fig. 2.7. Rates of reaction of *trans*-[$Pt(py)_2Cl_2$] in methanol at 30° as a function of the concentrations of different nucleophiles. From U. Belluco, L. Cattalini, F. Basolo, R. G. Pearson, and A. Turco, *J. Amer. Chem. Soc.*, 1965, **87**, 241.

acid has a large size, small charge, and has valence electrons which are easily distorted or removed (i.e., it is easily polarized); similarly a 'soft' base is one which is easily polarized. According to the Principle of Hard and Soft Acids and Bases,[35] hard acids prefer to coordinate with hard bases and soft acids prefer to coordinate with soft bases. It is to be expected then that platinum(II), a soft acid, should react rapidly with soft bases. A measure of the polarizability (or softness) of the reagent should be its ease of oxidation as indicated by the value of its oxidation potential (E°): the more easily polarizable the reagent the more positive the oxidation potential. The difficulty to this approach is the lack of accurate E° values since the couple Y–Y in the reaction

$$Y(aq) \rightarrow \tfrac{1}{2}(Y\text{–}Y)(aq) + e^-(aq)$$

is unstable, in some instances. A plot of log k_Y versus E°, for substitutions in *trans*-$[Pt(py)_2Cl_2]$ in methanol solution, is hardly convincing. Other unsuccessful attempts to find suitable correlations have been made, including the use of data on the wavelength of charge-transfer bands (an indication of the ease with which an electron is transferred from the ligand to the metal).

Recourse has been made to a more empirical line of attack. A high degree of success has been achieved by an adaptation of the method of Swain and Scott[36] which has proved effective for organic systems. This method depends upon the choice of a standard substrate to which the reactivities (indicated by the rate of substitution in this substrate) of the nucleophiles are referred. It is convenient to define, on the

Table 2.15

Values of nucleophilic reactivity constants, n_{Pt}^0, defined for various reagents Y by selecting *trans*-$[Pt(py)_2Cl_2]$ in methanol at 30° as a standard substrate*

Entering group, Y	n_{Pt}^0	Entering group, Y	n_{Pt}^0
CH_3OH	0	Br^-	3·96
CH_3O^-	<2·4	C_6H_5SH	4·14
Cl^-	3·04	I^-	5·42
NH_3	3·06	SCN^-	5·65
C_5H_5N	3·13	SO_3^{2-}	5·79
NO_2^-	3·22	$SeCN^-$	7·10
N_3^-	3·58	$C_6H_5S^-$	7·17
NH_2OH	3·85	Thiourea	7·17
N_2H_4	3·86	$S_2O_3^{2-}$	7·34

* From U. Belluco, in *Coordination Chemistry Reviews*, Vol. 1, p. 111, Elsevier, 1966.

basis of the substrate *trans*-[$Pt(py)_2Cl_2$], a *nucleophilic reactivity constant*, $n_{Pt}{}^0$, by the equation

$$\log\left(\frac{k_Y}{k_Y{}^0}\right)_0 = n_{Pt}{}^0$$

where k_Y is the second-order rate constant for the substitution of Y in the reference substrate in methanol at 30°, and $k_Y{}^0$ is the second-order

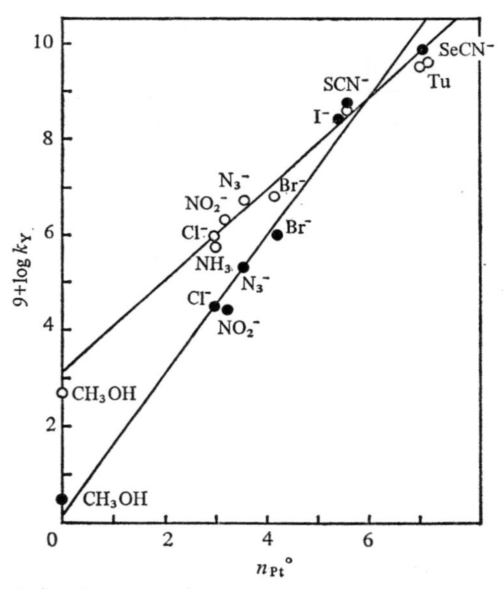

Fig. 2.8. Correlation between the rates of reaction of Pt(II) complexes in methanol at 30° with the standard *trans*-[$Pt(py)_2Cl_2$] for different nucleophiles: *trans*-[$Pt(PEt_3)_2Cl_2$] denoted by ●; *trans*-[$Pt(pip)_2Cl_2$] denoted by ○. From U. Belluco, in *Coordination Chemistry Reviews*, Vol. 1, p. 113, Elsevier, 1966.

rate constant for the corresponding solvent path ($k_Y{}^0 = k_S/[MeOH]$). Values of $n_{Pt}{}^0$ are given in Table 2.15. Linear plots are obtained (Fig. 2.8) of log k_Y for other Pt(II) complexes against $n_{Pt}{}^0$. Accordingly the relationship between log k_Y and $n_{Pt}{}^0$ can be written

$$\log k_Y = sn_{Pt}{}^0 + \log k_Y{}^0$$

since the intercept of a log k_Y–$n_{Pt}{}^0$ plot corresponds to a value of $n_{Pt}{}^0 = 0$ which occurs when $k_Y = k_Y{}^0$. The intercepts are thus a measure of the rate of reaction of the substrate with the solvent (the so-called *intrinsic reactivity*), and the slopes of the plots are equal to s, the *nucleophilic discrimination factor*, a characteristic of the complex which reflects the sensitivity of the substrate to alterations in the nature of the reagent.

Table 2.16 lists values of log $k_Y{}^0$ and s for a variety of Pt(II) complexes which have been studied in methanol. It is seen that complexes associated with small values of $k_Y{}^0$ (e.g., *trans*-[Pt$(PEt_3)_2Cl_2$] and

Table 2.16

Nucleophilic discrimination factors (s) and intrinsic reactivities (log $k_Y{}^0$) for several platinum(II) complexes in methanol*

Complex	log $k_Y{}^0$	s
trans-[Pt$(PEt_3)_2Cl_2$]	−8·82	1·43
trans-[Pt$(AsEt_3)_2Cl_2$]	−7·49	1·25
trans-[Pt(pip)$(PEt_3)Cl_2$]†	−7·35	1·15
trans-[Pt$(SeEt_2)_2Cl_2$]	−6·13	1·05
trans-[Pt$(pip)_2Cl_2$]	−5·82	0·91
trans-[Pt$(S(s\text{-}Bu)_2)_2Cl_2$]§	−5·74	0·57

* From U. Belluco, in *Coordination Chemistry Reviews*, Vol. 1, p. 112, Elsevier, 1966.
† pip = piperidine.
§ s-Bu = s-butyl.

trans-[Pt$(AsEt_3)_2Cl_2$] display large s factors. Conversely, those complexes having large $k_Y{}^0$ values possess small s values and are relatively insensitive to changes in the nucleophile. It is noteworthy that the triethylarsine and triethylphosphine complexes are those in which the transition state can be stabilized by π-bonding from metal to ligand. In general, regardless of the charge and nature of the complex, the reagents H_2O, Cl^-, pyridine, $N_3{}^-$, Br^-, I^-, SCN^-, $SO_3{}^{2-}$, and $S_2O_3{}^{2-}$ conform to the nucleophilicity scale ($n_{Pt}{}^0$ increasing from H_2O to $S_2O_3{}^{2-}$), but $NO_2{}^-$, $SeCN^-$, and thiourea (tu) behave anomalously.[37] These latter reagents are classed as *biphilic*, that is, in addition to σ-bonding, π-bonding occurs from metal to reagent. Thus ligands which increase the electron density on the metal impart greater reactivity to the biphilic reagent. Likewise, a biphilic reagent shows an abnormally high reactivity towards anionic complexes (e.g., $[PtCl_4]^{2-}$).

Effect of non-labile group

From an awareness of the relative reactivities of ligands in square-planar platinum(II) complexes an important general principle has emerged which is known as the *trans effect*. In the complex

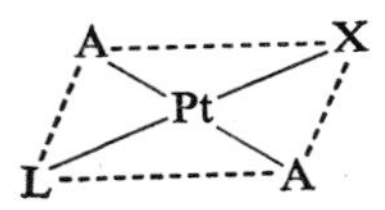

the ligand L affects the lability of the group X in the *trans* position. In this sense the *trans* effect is defined as the influence a non-labile group has upon the *rate* of replacement of the ligand opposite to it. Thus a ligand, L, is said to display a large *trans* effect if the reaction

A----X, Pt, L----A + Y ⟶ A----Y, Pt, L----A + X

is rapid and a small *trans* effect if the reaction is slow. The preparation of the *cis* and *trans* forms of $[Pt(NH_3)_2Cl_2]$ serves to illustrate this effect. The *trans* form is conveniently prepared by heating the tetra-ammine

H_3N----NH_3, Pt, H_3N----NH_3 $\xrightarrow[Cl^-]{\Delta}$

H_3N----Cl, Pt, H_3N----NH_3 $\xrightarrow[Cl^-]{\Delta}$ H_3N----Cl, Pt, Cl----NH_3

showing, in the second stage, that the ammonia opposite to chloride is more easily replaced than the ammonia opposite to ammonia. This indicates the order of labilization of the *trans* group to be $Cl^- > NH_3$. The *cis* isomer, on the other hand, is prepared by reaction of the tetrachloro complex with ammonia

Cl----Cl, Pt, Cl----Cl $\xrightarrow{NH_3}$ Cl----NH_3, Pt, Cl----Cl $\xrightarrow{NH_3}$ Cl----NH_3, Pt, Cl----NH_3

showing that the chloride opposite chloride is more easily replaced than the chloride opposite to ammonia, an observation consistent with the *trans* effect order $Cl^- > NH_3$.

From similar considerations of the stages involved in the syntheses of a wide range of platinum(II) complexes the general order of *trans* effect has been extended to

$$CN^- \sim C_2H_4 \sim CO \sim NO > SC(NH_2)_2 \sim SR_2 \sim PR_3 \sim SO_3H^- \sim NO_2^- \sim I^- \sim CNS^- > Br^- > Cl^- > py > RNH_2 \sim NH_3 > OH^- > H_2O$$

This series has been of immense value in synthetic work. For example, recognizing that the *trans* effect order is $NO_2^- > Cl^- > NH_3$, the synthesis of *cis*- and *trans*-$[PtNH_3NO_2Cl_2]$ can be performed simply and elegantly by reversing the order of substituting groups

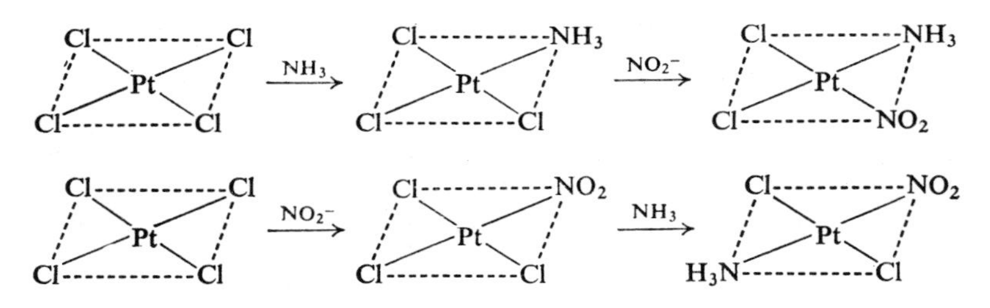

As a further example, the three possible geometrical isomers of $[PtNH_3pyClBr]$ can be prepared by observing the order $Br^- > Cl^- > py > NH_3$ (reactions (1) to (3), p. 69).

It should be noted that, in the third stage of the first synthesis, chloride is replaced in preference to ammonia, indicating that the strength of the metal–ligand bond has to be considered as well as the labilizing influence of the *trans* group. In this respect, it should be emphasized that the *trans* effect order represents nothing more than an average pattern of behaviour and, as such, should not be interpreted too literally.

Some quantitative studies have been made of the *trans*-labilizing influence of a limited number of ligands in platinum(II) complexes. Typical of these is an examination of the rate of substitution of pyridine in complexes of the type *trans*-$[PtA_2LCl]$ in ethanol solution, where $A = PEt_3$, and L is the non-labile ligand in the *trans* position to the replaced chloride group.[32] An inspection of the k_S and k_Y values reveals the *trans* effect order to be $PMe_3 > PEt_3$

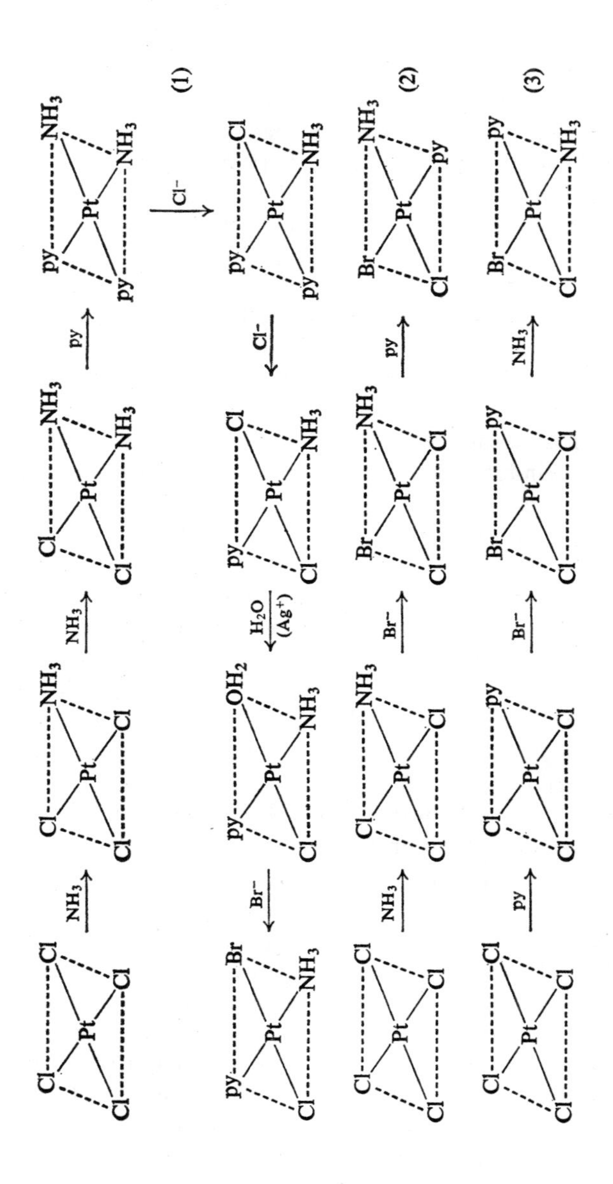
NH$_3$
NH$_3$
py
(1)
Br$^-$
H$_2$O (Ag$^+$)
Cl$^-$
Cl$^-$
NH$_3$
Br$^-$
py
(2)
py
Br$^-$
NH$_3$
(3)

$\sim H^- > PR_3 > CH_3^- > C_6H_5^- >$ *o*-tolyl > mesityl > Cl^-. In particular the ratio of the rates for hydride, methide, phenyl, and chloride, as the *trans* group, is 100,000:200:30:1, respectively. When the groups are in the *cis* position (as in *cis*-$[PtA_2LCl]$) the differences in reactivity are far less pronounced (methide:phenyl:chloride is 3·6:2·3:1). A separate kinetic investigation[38] of $[PtNH_3LCl_2]$ complexes has yielded a *trans* effect order of $C_2H_4 > NO_2^- > Br^- > Cl^-$ in the approximate ratio > 100:9:3:1 (as given by the k_Y values; the activation energies for NO_2^-, Br^-, and Cl^- were 11, 17, and 19 kcal mole^{-1}, respectively).

Two general theories of the *trans* effect are currently held: the polarization theory (of Grinberg[39]) and the π-bonding theory (of Chatt[40] and of Orgel[41]). The polarization theory considers the effect to be principally electrostatic in origin. When the complex contains four identical ligands, each ligand is polarized by the metal ion to the same extent and no induced dipole results in the central metal ion. However, when one of the ligands (L) is more polarizable than the others an induced dipole on the central metal ion results, the electron density is distorted via a σ-bond towards the *trans* position and the metal–X bond becomes lengthened and weakened in the transition state. Thus there should be a direct relationship between the polarizability of the *trans* ligand and its labilizing ability. Also, since the principal requirement is a metal ion which itself is polarizable, the effect should be less evident in palladium(II) complexes (a prediction borne out in practice). Although subject to some criticism[42], the polarization theory appears to account nicely for the behaviour of such ligands as hydride, methide, phenyl, and chloride where π-bonding is expected to be of minor importance. A number of observations of different types lend support. Measurements on *trans*-$[Pt(PEt_3)_2LCl]$ complexes reveal a decrease in dipole moment as L is varied from H^- to CH_3^- to phenyl to Cl^-. As determined by X-ray diffraction, the Pt—Br bond length in $[Pt(PEt_3)_2HBr]$ (where the bromide is *trans* to the highly polarizable hydride ion) is significantly greater than the sum of the atomic radii. Furthermore, for *trans*-$[Pt(PEt_3)_2HX]$, the infrared frequency of the Pt—H bond (and therefore the Pt—H bond strength) decreases in the order $X = NO_3^- > Cl^- > Br^- > I^- > NO_2^- > SCN^- > CN^-$, which is the exact reverse of the *trans* effect order of X.

The alternative theory explains the *trans* effect of groups such as C_2H_4, CN^-, CO, thiourea, and PR_3 in terms of their ability to π-bond with the metal. In these cases a reduction of the electron density on the metal, as a result of back donation from metal to ligand, facilitates the addition of an incoming group via a trigonal bipyramidal transition state. Moreover, since in the transition state the *trans* ligand is in the same trigonal plane as the entering and leaving groups, the transition state is stabilized with respect to the ground state. Therefore the difference in energy between ground and transition states (the activation energy) is less than it would be if no π-bonding occurred. Thus π-acceptor ligands occupy a high position in the *trans* series. In conclusion, some ions which can function as σ-donors and/or π-acceptors (e.g., iodide) owe their marked labilizing ability to both polarization and π-bonding influences.

Effect of solvent

A detailed investigation has been made[43] of the effect of solvent on the chloride exchange of *trans*-$[Pt(py)_2Cl_2]$

$$trans\text{-}[Pt(py)_2Cl_2] + \overset{*}{Cl}^- \rightarrow trans\text{-}[Pt(py)_2\overset{*}{Cl}_2] + Cl^-$$

In aqueous solution the rate of this reaction is virtually independent of chloride ion concentration and the exchange proceeds almost completely through the solvent path (k_S). Similar behaviour is noted when the reaction is conducted in the solvents dimethylsulphoxide (DMS), nitromethane, ethanol, and 1-propanol. The results reproduced in Table 2.17 show that the solvent effect on the value of k_S is in the order DMS > H_2O, $MeNO_2$ > ROH. A separate study of the rate of exchange of *trans*-$[Pt(PEt_3)_2Cl_2]$ shows also that DMS $\gg$ ROH. There is no parallel between the dielectric constant or general solvating properties of the solvent and the rate of exchange: in fact, taking into account these factors, water should be more effective than alcohols and the latter more effective than nitromethane. The unexpected behaviour of nitromethane as a solvent has been explained on the grounds of its ability to form π-bonds with the filled d_{yz} or d_{xz} orbitals of the platinum atom, thus stabilizing the transition state and permitting the displacement of the chloride ion: in this sense such solvents are termed biphilic. Contrasting behaviour is displayed by a number of solvents which, possessing poor coordinating power towards Pt(II), contribute little to a k_S path, exchange proceeding principally through a reagent-dependent path (k_Y). In this class the order of reactivity is carbon tetrachloride > benzene > *m*-cresol > t-butanol > ethyl acetate > acetone > dimethylformamide (DMF) (Table 2.17).

The nature of the solvent (acetone, DMS, and methanol) does not affect the nucleophilicity order for *trans*-$[Pt(PEt_3)_2Cl_2]$.[44] On the other hand, the reactivity order for saturated carbon substrates (e.g.,

Table 2.17

Effect of solvent on the chloride exchange of *trans*-$[Pt(py)_2Cl_2]$ at 25°*

Solvents in which exchange proceeds via the solvent path	$10^5 k_S$, s^{-1}
DMS	38
H_2O	3·5
$MeNO_2$	3·2
EtOH	1·4
n-C_3H_7OH	0·42
Solvents in which exchange proceeds via the reagent path	k_{Cl^-}, M^{-1} s^{-1}
CCl_4	10^4
C_6H_6	10^2
m-cresol	10^{-1}
t-C_4H_9OH	10^{-1}
EtOAc	10^{-2}
$(CH_3)_2CO$	10^{-2}
DMF	10^{-3}

* From ref. (43).

alkyl halides) is reversed in going from protic to aprotic solvents. The difference is, essentially, that solvation effects override the polarizability factor in the case of carbon whereas for reactions at soft centres, like platinum, solvation plays a secondary role and polarizability is all-important.

Mechanism of substitution

The essential requirement for a first-order solvent path, involving an aquo-complex intermediate, is that the rate of reaction of the intermediate with an incoming ligand should be in excess of the rate of formation of the intermediate. This is shown to be the case in the substitution reactions of the complex $[Pt(dien)X]^+$ where it has been possible to study separately the kinetics of the postulated intermediate $[Pt(dien)H_2O]^{2+}$ with a number of reagents including Cl^-, I^-, NO_2^-, SCN^-, OH^-, and pyridine.[45] From the data of Table 2.18 it is seen

Table 2.18

Rates of reaction of $[Pt(dien)H_2O]^{2+}$ with various entering groups (Y) in aqueous solution at 25°*

Y	[*Y*], M	$10^2 k_{obs}$, s^{-1}	Reactivity† relative to $[Pt(dien)Br]^+$
OH^-	0·005	very large	very large
I^-	0·001	3·9	50
SCN^-	0·001	1·45	20
Cl^-	0·005	0·53	50
NO_2^-	0·0009	0·090	30
pyridine	0·005	0·19	100

* From ref. (45).
† Expressed as ratio of k_{obs} for aquo complex to k_{obs} for bromo complex, for same initial concentrations of reactants.

that, in each instance, the aquo complex is much more reactive than the parent complex. Also, the order of reactivity towards the aquo complex is $OH^- \gg I^- > SCN^- > Cl^- > NO_2^- >$ py, comparable (with the important exception of OH^-) to the order observed for substitution reactions of $[Pt(dien)X]^+$. The extremely high reactivity of the hydroxide ion towards $[Pt(dien)H_2O]^{2+}$ is to be expected since reaction to $[Pt(dien)OH]^+$ requires only the transfer of a proton. Three possible mechanisms have been suggested for the solvent path, leading to an intermediate aquo complex (which is assumed to have a tetragonal structure):

(*a*) an associative-type mechanism in which solvent participates and a trigonal bipyramidal transition state is involved;
(*b*) a purely dissociative-type mechanism without solvent intervention in the transition state;

(*c*) a dissociative process in which solvent participates and a square-pyramidal transition state is formed.

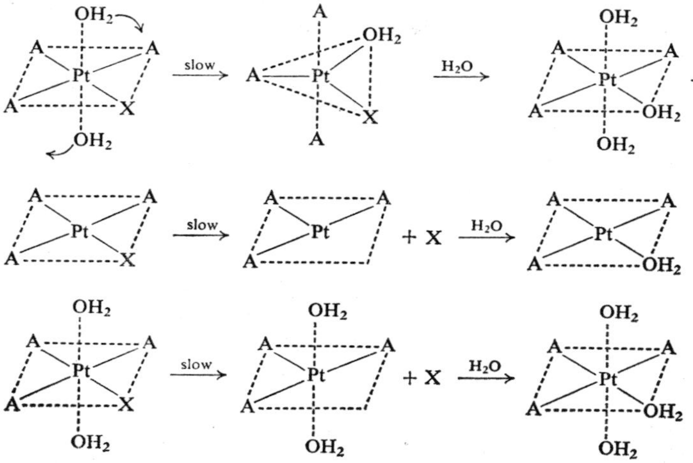

$PtA_3X(OH_2)_2$ $\xrightarrow{\text{slow}}$ $[A_3Pt(OH_2)X]$ $\xrightarrow{H_2O}$ $PtA_3(OH_2)(OH_2)_2$ + X (1)

PtA_3X $\xrightarrow{\text{slow}}$ PtA_3 + X $\xrightarrow{H_2O}$ PtA_3OH_2 (2)

$PtA_3X(OH_2)_2$ $\xrightarrow{\text{slow}}$ $PtA_3(OH_2)_2$ + X $\xrightarrow{H_2O}$ $PtA_3(OH_2)(OH_2)_2$ (3)

As we have seen the evidence against scheme (2) is three-fold: (*a*) the rate constant (k_S) is unaffected by changes in the net charge of the complex, (*b*) the values of k_S decrease with increase in steric hindrance, and (*c*) the values of k_S in different solvents bear no relationship to the dielectric constant of the medium but show a correlation with the coordinating power of the solvent. The reactions of $[Pt(dien)H_2O]^{2+}$ with chloride and nitrite follow the normal rate law: plots of k_{obs} versus $[Y]$ are linear with a common intercept. The implication of the latter observation is that a first-order path, independent of the concentration of the reagent, is available for $[Pt(dien)H_2O]^{2+}$: the only possibility is dissociation to give a transition state of square pyramidal structure. It follows that the postulation of a similar square pyramidal structure in scheme (3) can have no validity. From this evidence alone it would appear that scheme (3) is eliminated and that a trigonal bipyramid is the only reasonable transition state for the solvent path (as given in scheme (1)).

The same systems have been employed in an ingenious manner to obtain direct information on the mechanism of the reagent-dependent path (k_Y).[46] As well as the obvious reaction via a trigonal bipyramidal transition state (scheme (4)), there is an alternative mechanism in

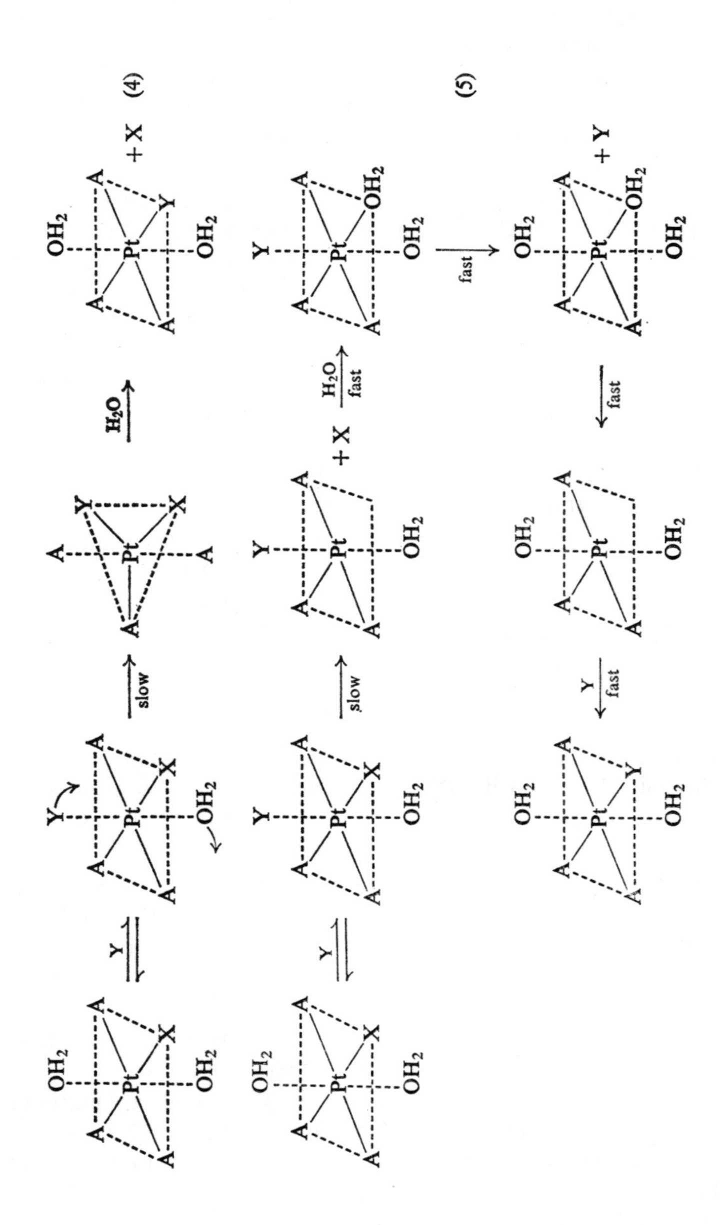
slow
H₂O
(4)
slow
fast
(5)
fast
fast

which the transition state is a square pyramid from which an aquo complex is formed by dissociation of a ligand (scheme (5)).

Both mechanisms have essentially associative rate-determining steps; however, no aquo-complex intermediate is formed in scheme (4). It will be recalled that OH^- reacts very rapidly (and quantitatively) with the aquo-complex, $[Pt(dien)H_2O]^{2+}$, to generate $[Pt(dien)OH]^+$ but is unreactive towards the parent complex, $[Pt(dien)X]^+$. Thus it has proved possible to devise a competition experiment allowing the detection of $[Pt(dien)H_2O]^{2+}$ if it were an intermediate in the reaction of $[Pt(dien)X]^+$ with a reagent, Y (scheme (5)): Y was allowed to react with $[Pt(dien)X]^+$ in the presence of OH^- ion. The results showed conclusively that the first step in the reaction is the formation of $[Pt(dien)Y]^+$ at a rate identical with its formation in the absence of OH^-. Moreover, *no* $[Pt(dien)OH]^+$ is formed along this route. This primary step is followed by a much slower conversion of $[Pt(dien)Y]^+$ to $[Pt(dien)OH]^+$ which proceeds at the normal rate expected for the k_S path. It is certain therefore that square-planar substitution does not occur through processes of the type depicted in (5). However, the alternative trigonal bipyramid mechanism, shown in (4), is acceptable. It has been suggested that the vacant $6p_z$ orbital of Pt(II) is utilized in forming the five-coordinated intermediate.

Effect of leaving group

The effect of replaced ligand X on the rates of substitution of the same nucleophile Y in $[Pt(dien)X]^+$ complexes has been the subject of a number of investigations.[47] The most recent study[48] shows that the ease of replacement of X follows the order $Cl^- \sim Br^- \sim I^- \gg N_3^- > SCN^- > NO_2^- > CN^-$ (high rates of reaction paralleling low enthalpies of activation). In other words, the ligands Cl^-, Br^-, and I^- are displaced with comparable ease whereas N_3^-, SCN^-, NO_2^-, and CN^- are more difficult to replace and, moreover, the rate of replacement is dependent on the nature of the ligand. At first thought, the behaviour of the halide groups is remarkable since the strength of the metal–halide bond increases in the order $Cl^- < Br^- < I^-$.

However, the observations can be given a rational explanation on the basis of the reaction profiles reproduced in Fig. 2.9. For each reaction profile there are two possible unsymmetrical transition states (A) and (B) of different relative energies along with an unstable five-coordinated symmetrical intermediate [Y—Pt—X]. In Fig. 2.9(a) the activation energy for the reaction is given by the energy difference between the ground state and the higher energy transition state which is A. In the formation of the activated complex, bond fission of the Pt—X bond is of relative unimportance compared with bond formation to the reagent Y. Consequently the rate of reaction should not

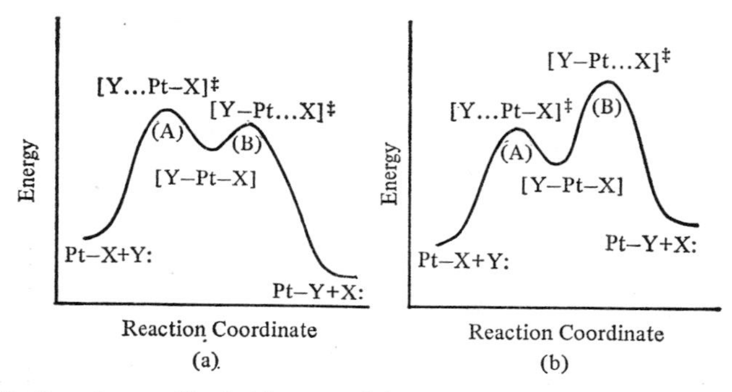

Fig. 2.9. Reaction profile (arbitrary scale) for associative mechanism involving an unstable five-coordinated intermediate in the displacements on $[Pt(dien)X]^+$. (*a*) For X = Cl^-, Br^-, I^-, the Pt—X bond rupture is kinetically unimportant. (*b*) For X = N_3^-, NO_2^-, SCN^-, CN^-, the bond rupture of Pt—X is kinetically important. From U. Belluco, R. Ettorre, F. Basolo, R. G. Pearson, and A. Turco, *Inorg. Chem.*, 1966, **5**, 591

depend upon the nature of X: such is the case for Cl^-, Br^-, and I^-. On the other hand, if the intimate path of reaction conforms to the profile given in Fig. 2.9(b), then the obstacle to reaction is the energy gap between ground state and transition state (B) where breaking of the Pt—X bond is of prime importance. Thus the rate of replacement of X should depend on the nature of X: such is the case for N_3^-, SCN^-, NO_2^-, and CN^-.

Electrophilic catalysis

Additional evidence for a five-coordinated, unsymmetrical transition state is provided by a report that the reactions of *trans*-$[Pt(pip)_2Cl_2]$ (where pip = piperidine) with $^{36}Cl^-$ and NO_2^- in methanol solution are catalysed by boric and nitrous acids.[49] A three-term rate law applies

$$k_{obs} = k_S + k_Y[Y^-] + k_C[HA][Y^-]$$

where k_S and k_Y have the usual significance, and k_C is the rate constant for the path in which complex, reagent Y^-, and acid HA are present together in the activated complex. The suggested mechanism for the catalysis is similar to that given in scheme (4), p. 75, except that catalyst and solvent share the axial sites above and below the plane; the formation of the trigonal bipyramidal structure is assumed to be rate-determining

CH_3OH / Cl, pip, Pt, pip, Cl / CH_3OH $\rightleftharpoons$ HNO_2 / Cl, pip, Pt, pip, Cl / CH_3OH $\rightleftharpoons$ HNO_2 / Cl, pip, Pt, pip, Cl / Y

$\downarrow$ slow

pip / Y, Cl, Pt, Cl / pip $\longrightarrow$ CH_3OH / Cl, pip, Pt, pip, Y / CH_3OH

It is noteworthy that the symmetrical five-coordinated species is considered to be an active intermediate and not an activated complex, since the rate expression shows that the catalyst must be present in the transition state. Possibly the function of the catalyst is to bind to the complex in some way and thereby stabilize the (unsymmetrical) transition state and allow easier entry of the reagent. This could be brought about if the catalyst were to function as an electrophile, removing π electrons from the Pt(II) reaction centre and thus assisting the metal in accepting σ electrons from the nucleophile. In this respect it is interesting that the reactions of *trans* complexes of the type $[Pt(PEt_3)_2Cl_2]$, $[Pt(AsEt_3)_2Cl_2]$, and $[Pt(pip)_2(NO_2)Cl]$ are not catalysed by nitrous acid. Here, the complexes already contain π-electron withdrawing ligands (PEt_3, $AsEt_3$, and NO_2^-, respectively) and the increased electron density in the transition state, caused by attachment of the nucleophile, can be alleviated without the need for an external electrophile. It is also of interest that the reactions of nitrite ion with *trans*-$[Pt(NH_3)_2Cl_2]$ and $[PtCl_4]^{2-}$ are susceptible to catalysis by small amounts of nitrous acid formed by the hydrolysis of NO_2^-.

Nickel(II) and palladium(II) complexes

There is a general paucity of information on the substitution reactions of palladium(II) complexes but sufficient work has been done to expose the main features. Analogous palladium(II), platinum(II), and nickel(II) complexes of the triethylphosphine type have been studied in relation to their reaction with pyridine in ethanol

$$[M(PEt_3)_2RCl] + py \rightleftharpoons [M(PEt_3)_2Rpy]^+ + Cl^-$$

The kinetic results of Table 2.19 show that the rates of substitution of

Table 2.19

Rates of reaction* of some analogous Ni(II), Pd(II), and Pt(II) complexes at 25°†

Complex	k_s, s^{-1}
trans-$[Ni(PEt_3)_2(o\text{-tolyl})Cl]$	33
trans-$[Pd(PEt_3)_2(o\text{-tolyl})Cl]$	0·58
trans-$[Pt(PEt_3)_2(o\text{-tolyl})Cl]$	$6{\cdot}7 \times 10^{-6}$
trans-$[Ni(PEt_3)_2(mesityl)Cl]$	0·02
trans-$[Pt(PEt_3)_2(mesityl)Cl]$	$1{\cdot}2 \times 10^{-6}$

* From ref. (32).

† Substitution of pyridine for chloride in ethanol solution.

trans-$[M(PEt_3)_2(o\text{-tolyl})Cl]$, where M = Ni, Pd, and Pt, are in the ratio of approximately $5 \times 10^6 : 10^5 : 1$, respectively. The extreme variation in reactivity between Pt(II) and Ni(II) complexes is explainable in terms of the mechanism described by scheme (4) on page 75, in which a ligand above the plane of the square moves in to displace the halide ion, since an expansion of coordination number is easier for Ni(II) than for Pt(II). When the above-plane position is partially blocked, as in the complexes *trans*-$[M(PEt_3)_2(mesityl)Cl]$, then the difference in reactivity of Ni(II) and Pt(II) is less pronounced, the Ni(II) complex reacting only about 20,000 times faster than the Pt(II) complex.

The reactions of $[Pd(dien)Cl]^+$ with hydroxide and bromide ions are too fast to be measured by the stopped-flow method ($t_{1/2} < 10^{-3}$ s) even at concentrations as low as 10^{-3} M. However, the sterically-hindered $[Pd(Et_4dien)Cl]^+$ complex (where $Et_4dien = (C_2H_5)_2NCH_2CH_2NHCH_2CH_2N(C_2H_5)_2$) reacts with various nucleophiles at 25° with a half-life of about 6 minutes.[50] This substrate, in which the four ethyl groups completely block the space above and below the planar complex, takes on an octahedral appearance and for this reason is referred to as *pseudo-octahedral*. The reaction of $[Pd(Et_4dien)Cl]^+$ (and similar complexes) is independent of the concentration of reagent for the case of Br^- and I^- but not for OH^- ion. The dependence of the rate on hydroxide-ion concentration is most untypical of square-planar substitutions and is reminiscent of the behaviour of chloroamminecobalt(III) complexes (see p. 41). Since the normal associative mechanism seems unlikely, it is reasonable to postulate that a conjugate-base mechanism (p. 42) operates due to the presence of a N—H hydrogen

$$[Pd(Et_4dien)Cl]^+ + OH^- \overset{K}{\rightleftharpoons} [Pd(Et_4dien-H)Cl] + H_2O$$

$$[Pd(Et_4dien-H)Cl] \xrightarrow[k_{CB}]{slow} [Pd(Et_4dien-H)]^+ + Cl^-$$

$$[Pd(Et_4dien-H)]^+ + H_2O \xrightarrow{fast} [Pd(Et_4dien)OH]^+$$

$$[Pd(Et_4dien)Cl]^+ + H_2O \xrightarrow[k_S]{slow} [Pd(Et_4dien)H_2O]^{2+} + Cl^-$$

$$[Pd(Et_4dien)H_2O]^{2+} + OH^- \xrightarrow{fast} [Pd(Et_4dien)OH]^+ + H_2O$$

The rate law is then derived as

$$\text{rate} = k_S[complex] + k_{CB}\,K[complex][OH^-]$$

where $K = K_A/K_W$ (K_A is the acid dissociation constant of the complex). Estimated values of k_S and k_{CB} showed that base hydrolysis is about 30 times more effective than acid hydrolysis.[51] This factor is similar to that obtained[52] for the analogous gold(III) complex, $[Au(Et_4dien)Cl]^{2+}$. The conjugate-base scheme is supported by the observation that hydroxide ion has no effect on the rate of reaction of $[Pd(MeEt_4dien)Cl]^+$, a complex which contains no N—H hydrogen ($MeEt_4dien = (C_2H_5)_2NCH_2CH_2N(CH_3)CH_2CH_2N(C_2H_5)_2$).

Substitution reactions of bis-(acetylacetonate)palladium(II) with a series of nucleophiles in a water–methanol solvent have been studied spectrophotometrically[53]

$$[Pd(acac)_2] + 4Y^- + 2H^+ \rightarrow [PdY_4]^{2-} + 2Hacac$$

As might be expected, the kinetic data display the features of an associative mechanism (with the added complication that the rate shows a pH-dependence since the leaving group is a chelate). The sequence of nucleophilic reactivity is $SCN^- > I^- > Br^- > Cl^- > OH^- \sim H_2O$ which is identical with the order of reactivity in Pt(II) complexes: again hydroxide ion is revealed as an ineffective reagent. Seemingly, as with Pt(II), polarizability is the dominant factor in Pd(II) reactions.

Gold(III) complexes

Scant attention has been given to complexes of gold(III), due largely to their comparative rarity and ease of reduction. A comparison has been made,[54] however, of the isoelectronic and isostructural complexes $[Au(dien)Cl]^{2+}$ (and/or its conjugate base $[Au(dien-H)Cl]^+$) and $[Pt(dien)Cl]^+$. Although the rates of substitution of these complexes are intrinsically different, the order of nucleophilic reactivity towards the Au(III) and Pt(II) substrates is identical, i.e., $I^- > SCN^- > Br^- > N_3^- > OH^- \sim H_2O$. The relative unimportance of the solvent-dependent path (k_S) for the Au(III) complexes contrasts with the behaviour of Pt(II) where the k_S path is appreciable and, occasionally, dominant (e.g., the exchange of chloride ion with $[PtCl_4]^{2-}$ shows first-order kinetics whereas chloride exchange of $[AuCl_4]^-$ displays both first- and second-order kinetics). A possible reason for this is that, relative to Pt(II), the greater charge of Au(III) renders bond-making more important than bond-breaking, implying that, in the transition state, charge neutralization is of more importance than charge separation (see p. 20). Further observations support this idea: for example, $[Au(dien)Cl]^{2+}$ reacts faster than $[Au(dien-H)Cl]^+$ or $[AuCl_4]^-$.

Rhodium(I) complexes

Nuclear magnetic resonance studies have shown that no exchange occurs, during 5 hours at 100°, between the ethylene in bis-(ethylene)-π-cyclopentadienylrhodium(I), $[C_5H_5Rh(C_2H_4)_2]$, and deuterated ethylene.[55] However, exchange is rapid between C_2D_4 and bis-(ethylene)-rhodium(I)acetylacetonate, $[(acac)Rh(C_2H_4)_2]$. Structural differences in the two complexes account for this behaviour. Using a simple valence-bond approach, rhodium(I) ($4s^2 4p^6 4d^8$) will have acquired ten electrons by coordinating with a π-cyclopentadienyl group (six electrons) and two ethylene molecules (two electrons each). Thus the complex will have achieved the inert-gas configuration of xenon and be inert to exchange. In contrast, by coordination with acetylacetone (four electrons) and two ethylene molecules (two electrons each), Rh(I) will be two electrons short of this inert configuration. Accordingly it can accommodate a further two electrons from an olefin molecule and exchange by an associative mechanism

$$[(acac)Rh(C_2H_4)_2] + C_2D_4 \rightarrow [(acac)Rh(C_2H_4)_2(C_2D_4)] \rightarrow [(acac)Rh(C_2H_4)(C_2D_4)] + C_2H_4$$

It should be noted, however, that rapid exchange of carbon monoxide occurs in the complex $[\pi\text{-}C_5H_5Rh(CO)_2]$ although this has an inert-gas configuration. Steric, as well as electronic, factors must be considered.

The kinetics of the reaction between $[Rh(cy)(SbR_3)Cl]$, where cy = 1,5-cyclo-octadiene and R = *p*-tolyl, and various amines in acetone solution has some unusual features: the value of k_S was found to vary with the nature of the incoming amine, and significant spectral changes occurred on mixing the reactants.[56] These observations suggest a mechanism, complex + amine $\overset{\text{fast}}{\rightleftharpoons}$ intermediate $\xrightarrow{\text{slow}}$ products, where the intermediate has a five-coordinated structure. There is a similar implication in the report[57] that the rate of replacement of the second chloride in *cis*-$[Pt(NH_3)_2Cl_2]$ by nitrite ion is non-linear in nitrite concentration. This behaviour was explained in terms of a rapid pre-equilibrium

$$cis\text{-}[Pt(NH_3)_2NO_2Cl] + NO_2^- \overset{K}{\rightleftharpoons} [Pt(NH_3)_2(NO_2)_2Cl]^-$$

with the rate-controlling step as

$$[Pt(NH_3)_2(NO_2)_2Cl]^- \xrightarrow{k} cis\text{-}[Pt(NH_3)_2(NO_2)_2] + Cl^-$$

Furthermore, spectrophotometric measurements showed that the addition of NO_2^- to a solution of *cis*-$[Pt(NH_3)_2NO_2Cl]$ gave rise to a marked and immediate increase in absorbance which was attributed to the formation of the chlorodinitro intermediate. However, it is possible that the change in absorbance might be due to the presence of nitrous acid resulting from the hydrolysis of NO_2^-.

It is of great current interest that square-planar complexes of d^8 Pt(II), Ir(I), and Rh(I) show the ability to add on covalent molecules such as hydrogen, halogens, methyl iodide, and oxygen to form six-coordinated complexes of d^6 configuration. The oxidative addition of H_2, O_2, and CH_3I to the Ir(I) complex *trans*-$[IrX(CO)(P(C_6H_5)_3)_2]$ has been investigated kinetically.[58]

Substitution reactions of five-coordinated platinum(II) complexes

An examination of substitution in five-coordinated complexes merits attention since they represent an intermediate case between the four-coordinated and six-coordinated systems. Little work has been done on such complexes because of their rarity and kinetic inflexibility, but a detailed investigation,[59] making use of a stopped-flow method, has been carried out on the reactions of the trigonal bipyramidal complexes of platinum(II) with the tetradentate ligand tris-(*o*-diphenylarsinophenyl)arsine, (*o*-$Ph_2As.C_6H_4)_3As$, (QAS)

$$[Pt(QAS)X]^+ + Y^- \rightarrow [Pt(QAS)Y]^+ + X^-$$

The results in methanol solution, although somewhat complicated by ion-pair formation, conform to a second-order rate law and demonstrate clearly a six-coordinated transition state for the reactions. The order of reactivity of nucleophiles is $CN^- > S{=}C(NH_2)_2 > SCN^- \sim I^- \sim Ph_3P > N_3^- > NO_2^-$ which corresponds to the order observed for the planar complexes of Pt(II) (with the exception of triphenyl phosphine). Excluding the latter reagent, there is a linear relationship between log k_Y and $n_{Pt}{}^0$ (using the values of $n_{Pt}{}^0$ assigned for (Pt(II)) although the intercept is at variance with the observed value of k_S: this disparity is ascribed to steric hindrance in the transition state.

The difference in rate between $[Pt(QAS)Br]^+$ and $[Pd(QAS)Br]^+$ is comparable with that of the planar complexes of Pt(II) and Pd(II). Also the TAS complex of Pt(II) (TAS = bis-(*o*-diphenylarsinophenyl)-phenylarsine, (*o*-$Ph_2AsC_6H_4)_2AsPh$), which is square-planar, reacts about 10^4 times faster than the corresponding sterically-hindered QAS complex. It would appear that the five-coordinated systems are intermediate, in both a kinetic and steric sense, between planar complexes, which are open to nucleophilic attack, and octahedral complexes, which are closed to attack.

C. Tetrahedral complexes

Tetrahedral complexes are generally labile and because of this their substitution reactions have been little studied. In this section a brief account is given of the reaction mechanisms of compounds of the Group IV elements, silicon, germanium, and tin. Substitution reactions of metal carbonyls are also considered.

Substitution at silicon, germanium, and tin

Since carbon and silicon have similar electronic configurations (C, $1s^2 2s^2 2p^2$; Si, $1s^2 2s^2 2p^6 3s^2 3p^2 (3d^0)$), a close resemblance between the chemistry of analogous carbon and silicon compounds is to be expected. To a large extent this is observed in practice. However, in general, the reactivity of tetrahedral silicon compounds is much greater than that of their carbon analogues. For example, the hydrolysis of CH_3I

$$CH_3I + H_2O \rightarrow CH_3OH + HI$$

is slow whereas the hydrolysis of SiH_3I

$$SiH_3I + H_2O \rightarrow SiH_3OH + HI$$

is extremely rapid, although both compounds are unstable (in a thermodynamic sense) with respect to their products of hydrolysis. The greater reactivity (lability) of the silicon compound is ascribed to the ease with which silicon can form a five-coordinated intermediate in which the incoming reagent attaches itself to the silicon atom. The formation of such an intermediate (via a low-energy path) is more favourable in the case of silicon than in the case of carbon because (*a*) the silicon atom is much greater in size than the carbon atom (the atomic radii are 1·17 Å and 0·771 Å, respectively), and (*b*) silicon, having a vacant 3*d* orbital, unlike carbon, is more prone to expand its coordination number to greater than four. Certainly there is a lack of kinetic evidence to suggest that a dissociative mechanism is possible in the case of substitution at silicon.

The rates of hydrolysis of several triaryl halides of germanium, R_3GeX, have been investigated.[60] Since the polarity of the M—X bond tends to increase successively along the group IV elements in the sequence Si < Ge < Sn, it might be anticipated that the formation of an ion analogous to the carbonium ion should be more favoured in the case of germanium than for silicon. However, this prediction is shown to be false on the grounds that, for a particular solvent (dioxan–water or acetone–water), the rate decreases in the order $R_3GeBr > R_3GeCl > R_3GeF$. Furthermore, the decrease in rate on replacing R = phenyl by R = *p*-tolyl allows a discrimination to be made between

$$\begin{aligned} R_3Ge{-}X &\xrightarrow{\text{slow}} R_3Ge^+ + X^- \\ R_3Ge^+ + H_2O &\xrightarrow{\text{fast}} R_3Ge{-}OH_2{}^+ \end{aligned} \qquad (1)$$

and

$$\begin{aligned} R_3Ge{-}X + H_2O &\overset{\text{fast}}{\rightleftharpoons} R_3Ge\begin{matrix} \diagup X \\ \diagdown OH_2 \end{matrix} \\ R_3Ge\begin{matrix} \diagup X \\ \diagdown OH_2 \end{matrix} &\xrightarrow{\text{slow}} R_3Ge{-}OH_2{}^+ + X^- \end{aligned} \qquad (2)$$

If mechanism (1) were operative then $(p\text{-tolyl})_3GeF$ should hydrolyse faster than $(C_6H_5)_3GeF$ since the *p*-tolyl group is more electron-releasing than the phenyl group; however, electron release to germanium will result in inhibition if mechanism (2) holds.

The solvolyses of organotin halides of the type R_3SnCl (where R = ethyl, iso-propyl, t-butyl or phenyl) have been examined kinetically in ethanol, 2-propanol, and water–dioxan media.[61] The following significant feature emerges: as the size and inductive power of the R group increases, solvolysis becomes less favoured. This is contrary to the behaviour shown by alkyl halides and is incompatible with reaction via a stannonium ion, R_3Sn^+.

Reactions of metal carbonyls

The exchange reactions of carbon monoxide with carbonyls of the first-row transition elements show pronounced variations in rate. At room temperature, the exchanges of $Ni(CO)_4$, and $Co_2(CO)_8$ are rapid whereas those of $Fe(CO)_5$, $Cr(CO)_6$ and $Mn_2(CO)_{10}$ are extremely slow. This difference in reactivity is no doubt related to the coordination state of the carbonyl: the four-coordinated nickel carbonyl shows greater lability than the carbonyls of five-coordinated iron and six-coordinated chromium. There would appear to be some similarity here to the behaviour of cyanide complexes as regards their exchanges with CN^- (an ion which is isolectronic with CO): four-coordinated species like $[Ni(CN)_4]^{2-}$, $[Pd(CN)_4]^{2-}$, and $[Hg(CN)_4]^{2-}$ exchange rapidly in contrast to the slow exchanges of $[Cr(CN)_6]^{3-}$, $[Fe(CN)_6]^{3-}$, $[Fe(CN)_6]^{4-}$, and $[Co(CN)_6]^{3-}$. The exchange of $Ni(CO)_4$ in toluene solution with ^{14}C-labelled carbon monoxide has been investigated kinetically.[62] The fact that the rate of exchange is independent of the concentration of carbon monoxide suggests a dissociative route for the exchange

$$Ni(CO)_4 \rightleftharpoons Ni(CO)_3 + CO$$

The observed activation energy of 13 kcal mole^{-1} (obtained from rate data over the range $-14°$ to $25°$) agrees closely with the value obtained for the thermal dissociation of $Ni(CO)_4$ in the gas phase. A dissociative mechanism is to be expected on a number of counts. Firstly, the filled *d* orbitals of Ni(O), a d^{10} system, should discourage attack by an incoming ligand. Secondly, calculations show that an increase in π-bond character of the metal–ligand bond occurs on changing from a sp^3 tetrahedral to a sp^2 triangular planar structure, an effect which assists in stabilizing the transition state, $Ni(CO)_3$. Thirdly, estimated values show the average bond energy of the M—C bond to be less in $Ni(CO)_4$ (44 kcal) than in $Fe(CO)_5$ and $Cr(CO)_6$. It will be noted that this value of 44 kcal is considerably higher than the apparent energy of activation for carbon monoxide exchange. This can only be rationalized on the grounds that the first metal–ligand bond is broken more easily than the remaining bonds. It is pertinent that a comparison of the rates of exchange of triphenylphosphine derivatives discloses the order of reactivity $Ni(CO)_4 > Ni(CO)_3PR_3 > Ni(CO)_2(PR_3)_2$.

Recently, the rates of exchange and substitution of $Ni(CO)_4$ have been compared. Both ^{14}CO exchange and reaction with triphenylphosphine or trimethyl phosphite were found to be first order in $Ni(CO)_4$ and independent of the reagent concentration.[63] Although the rates of reaction at 0° in toluene roughly correspond, evaluation of the activation parameters reveals that the two processes are quite separate and that they take place by different mechanisms. The agreement in rates at 0° is fortuitous. Further kinetic investigations are required before a reasonable explanation can be advanced for these observations.

Substitution reactions of $Co(NO)(CO)_3$

Nitrosyltricarbonylcobalt(O), $Co(NO)(CO)_3$, is isostructural and isoelectronic with tetracarbonylnickel(O), $Ni(CO)_4$. Thus it might be expected that the two compounds should show similar behaviour in their substitution reactions. However, the rate of replacement of carbon monoxide in $Co(NO)(CO)_3$

$$Co(NO)(CO)_3 + L \rightarrow Co(NO)(CO)_2L + CO$$

depends upon the concentration and nature of the nucleophile, following a second-order rate law

$$\text{rate} = k[Co(NO)(CO)_3][L]$$

where L is PR_3, $P(OR)_3$, AsR_3 or pyridine in toluene, tetrahydrofuran, and nitromethane solvents at 25°.[64] This suggests that substitution proceeds by an associative mechanism in which the five-coordinated species probably takes on a trigonal-bipyramidal structure (analogous to that adopted by $Fe(CO)_5$). The difference in behaviour of $Co(NO)(CO)_3$ and $Ni(CO)_4$ would seem to rest in the specific properties of NO as a ligand. The latter, by acting as a three-electron donor to the metal, produces a formal negative charge on the Co. Because of this the Co-carbonyl bond should show greater π-character than does the Ni-carbonyl bond in $Ni(CO)_4$. In this respect the C—O stretching frequency in the infrared spectrum is higher in $Ni(CO)_4$ than in $Co(NO)(CO)_3$. Thus the strength of the Co—C bond is greater than that of the Ni—C bond and the rupture of the latter (by a first-order process) is more easily accomplished. A correlation between reagent reactivity and base strength of the nucleophile has been uncovered. Since soft bases (i.e., phosphines and phosphites) react faster than hard bases (i.e., arsines and pyridine) at the soft cobalt centre, a parallelism between reagent reactivity and polarizability should result. Unfortunately data on polarizability are not available.

References

1. C. H. Langford and H. B. Gray, *Ligand Substitution Processes*, p. 14, Benjamin, 1965.
2. F. Basolo and R. G. Pearson, *Mechanisms of Inorganic Reactions*, Second Edition, p. 145, Wiley, 1967.
3. R. H. Holyer, C. D. Hubbard, S. F. A. Kettle, and R. G. Wilkins, *Inorg. Chem.*, 1966, **5**, 622.

4. R. G. Pearson, C. R. Boston, and F. Basolo, *J. Amer. Chem. Soc.*, 1953, **75**, 3089.
5. F. Basolo, J. G. Bergmann, R. E. Meeker, and R. G. Pearson, *J. Amer. Chem. Soc.*, 1956, **78**, 2676.
6(a). M. E. Baldwin, S. C. Chan, and M. L. Tobe, *J. Chem. Soc.*, 1961, 4637; (b). R. G. Pearson, C. R. Boston, and F. Basolo, *J. Phys. Chem.*, 1955, **59**, 304; (c). C. K. Ingold, R. S. Nyholm, and M. L. Tobe, *Nature Lond.*, 1960, **187**, 477.
7. F. Monacelli, F. Basolo, and R. G. Pearson, *J. Inorg. Nucl. Chem.*, 1962, **24**, 1241.
8. A. Haim and W. K. Wilmarth, *Inorg. Chem.*, 1962, **1**, 583.
9. F. Basolo, J. C. Hayes, and H. M. Neumann, *J. Amer. Chem. Soc.*, 1953, **75**, 5102; 1954, **76**, 3807.
10. K. V. Krishnamurty and G. M. Harris, *J. Phys. Chem.*, 1960, **64**, 346.
11. C. W. Merideth, W. D. Mathews, and E. F. Orlemann, *Inorg. Chem.*, 1964, **3**, 320.
12. A. Haim and W. K. Wilmarth, *Inorg. Chem.*, 1962, **1**, 573.
13. R. J. Grassi, A. Haim, and W. K. Wilmarth, in *Advances in the Chemistry of the Coordination Compounds* (ed. S. Kirschner), Macmillan, 1961; R. J. Grassi, A. Haim, and W. K. Wilmarth, *Inorg. Chem.*, 1967, **6**, 237.
14. J. Halpern, R. A. Palmer, and L. M. Blakley, *J. Amer. Chem. Soc.*, 1966, **88**, 2877.
15. R. G. Pearson, R. E. Meeker, and F. Basolo, *J. Amer. Chem. Soc.*, 1956, **78**, 709.
16. D. D. Brown, C. K. Ingold, and R. S. Nyholm, *J. Chem. Soc.*, 1953, 2674.
17. F. J. Garrick, *Nature Lond.*, 1937, **139**, 507.
18. M. Green and H. Taube, *Inorg. Chem.*, 1963, **2**, 948.
19. R. B. Jordan and A. M. Sargeson, *Inorg. Chem.*, 1965, **4**, 433.
20. For a treatment of the stereochemistry of the base hydrolysis of Co(III) complexes, see R. G. Pearson and F. Basolo, *Inorg. Chem.*, 1965, **4**, 1522.
21. S. C. Chan and F. Leh, *J. Chem. Soc.* (A), 1966, 126.
22. S. C. Chan and F. Leh, *J. Chem. Soc.* (A), 1966, 129.
23. M. L. Tobe, *Sci. Progr.*, 1960, **48**, 483; S. C. Chan and M. L. Tobe, *J. Chem. Soc.*, 1962, 4351.
24. R. G. Pearson, P. M. Henry, J. G. Bergmann, and F. Basolo, *J. Amer. Chem. Soc.*, 1954, **76**, 5920.
25. F. Basolo, M. L. Morris, and R. G. Pearson, *Disc. Faraday Soc.*, 1960, **29**, 80.
26. H. R. Ellison, F. Basolo, and R. G. Pearson, *J. Amer. Chem. Soc.*, 1961, **83**, 3943.
27. D. D. Brown and C. K. Ingold, *J. Chem. Soc.*, 1953, 2674.
28. R. G. Pearson, H. H. Schmidtke, and F. Basolo, *J. Amer. Chem. Soc.*, 1960, **82**, 4434.
29. M. N. Hughes and M. L. Tobe, *J. Chem. Soc.*, 1965, 1204.

30. For reviews (and key references) see (*a*) M. Eigen, *Pure and Appl. Chem.*, 1963, **6**, 97; (*b*) E. F. Caldin, *Fast Reactions in Solution*, pp. 94–98 (ultrasonics) and pp. 255–258 (n.m.r.), Blackwell, 1964; (*c*) N. Sutin, *Ann. Rev. Phys. Chem.*, 1966, **17**, 120–127.
31. T. J. Conocchioli, G. H. Nancollas, and N. Sutin, *Inorg. Chem.*, 1966, **5**, 1.
32. F. Basolo, J. Chatt, H. B. Gray, R. G. Pearson, and B. L. Shaw, *J. Chem. Soc.*, 1961, 2207.
33. U. Belluco, L. Cattalini, F. Basolo, R. G. Pearson, and A. Turco, *J. Amer. Chem. Soc.*, 1965, **87**, 241.
34. For a review, see R. G. Pearson, *J. Amer. Chem. Soc.*, 1963, **85**, 3533.
35. R. G. Pearson, *Chem. in Britain*, 1967, **3**, 103.
36. C. G. Swain and C. B. Scott, *J. Amer. Chem. Soc.*, 1953, **75**, 141.
37. L. Cattalini, A. Orio, and M. Nicolini, *J. Amer. Chem. Soc.*, 1966, **88**, 5734.
38. O. E. Zvyagintsev and E. F. Karandasheva, *Doklady Akad. Nauk S.S.S.R.*, 1955, **101**, 93.
39. A. A. Grinberg, *Acta Physicochim. U.R.S.S.*, 1935, **3**, 573.
40. J. Chatt, L. A. Duncanson, and L. M. Venanzi, *J. Chem. Soc.*, 1955, 4456.
41. L. E. Orgel, *J. Inorg. Nucl. Chem.*, 1956, **2**, 137.
42. See, for example, F. Basolo and R. G. Pearson, in *Progress in Inorganic Chemistry* (ed. F. A. Cotton), Vol. 4, pp. 417–419, Interscience, 1962. A molecular orbital treatment has been advanced by C. H. Langford and H. B. Gray, see ref. 1, pp. 25–29.
43. R. G. Pearson, H. B. Gray, and F. Basolo, *J. Amer. Chem. Soc.*, 1960, **82**, 787.
44. U. Belluco, M. Martelli, and A. Orio, *Inorg. Chem.*, 1966, **5**, 582.
45. H. B. Gray and R. J. Olcott, *Inorg. Chem.*, 1962, **1**, 481.
46. See ref. 1, pp. 38–39.
47. See, for example, F. Basolo, H. B. Gray, and R. G. Pearson, *J. Amer. Chem. Soc.*, 1960, **82**, 4200.
48. U. Belluco, R. Ettorre, F. Basolo, R. G. Pearson, and A. Turco, *Inorg. Chem.*, 1966, **5**, 591.
49. U. Belluco, L. Cattalini, F. Basolo, R. G. Pearson, and A. Turco, *Inorg. Chem.*, 1965, **4**, 925.
50. W. H. Baddley and F. Basolo, *J. Amer. Chem. Soc.*, 1964, **86**, 2075.
51. W. H. Baddley and F. Basolo, *J. Amer. Chem. Soc.*, 1966, **88**, 2944.
52. C. F. Weick and F. Basolo, *Inorg. Chem.*, 1966, **5**, 576.
53. R. G. Pearson and D. A. Johnson, *J. Amer. Chem. Soc.*, 1964, **86**, 3983.
54. W. H. Baddley and F. Basolo, *Inorg. Chem.*, 1964, **3**, 1087.
55. R. Cramer, *J. Amer. Chem. Soc.*, 1964, **86**, 217.
56. L. Cattalini, A. Orio, R. Ugo, and F. Bonati, *Chem. Commun.*, 1967, 48.
57. P. Haake, *Proc. Chem. Soc.*, 1962, 278.
58. P. B. Chock and J. Halpern, *J. Amer. Chem. Soc.*, 1966, **88**, 3511.
59. R. G. Pearson, M. M. Muir, and L. M. Venanzi, *J. Chem. Soc.*, 1965, 5521.

60. O. H. Johnson and E. A. Schmall, *J. Amer. Chem. Soc.*, 1958, **80**, 2931.
61. R. H. Prince, *J. Chem. Soc.*, 1959, 1783.
62. F. Basolo and A. Wojcicki, *J. Amer. Chem. Soc.*, 1961, **83**, 520.
63. L. R. Kangas, R. F. Heck, P. M. Henry, S. Breitschaft, E. M. Thorsteinson, and F. Basolo, *J. Amer. Chem. Soc.*, 1966, **88**, 2334.
64. E. M. Thorsteinson and F. Basolo, *J. Amer. Chem. Soc.*, 1966, **88**, 3929.

Bibliography

C. H. Langford and H. B. Gray, *Ligand Substitution Processes*, Benjamin, 1965.

F. Basolo and R. G. Pearson, Mechanisms of Substitution Reactions of Metal Complexes, in *Advances in Inorganic Chemistry and Radiochemistry* (ed. H. J. Emeléus and A. G. Sharpe), Vol. 3, p. 1, Academic Press, 1961.

R. G. Wilkins, Kinetics and Mechanism of Replacement Reactions of Coordination Compounds, in *Quarterly Reviews*, 1962, **16**, 316.

M. L. Tobe, The Mechanisms of Some Substitution Reactions of Octahedral Coordination Complexes, in *Science Progress*, 1960, **48**, 483.

D. R. Stranks, The Reaction Rates of Transitional Metal Complexes, in *Modern Coordination Chemistry* (ed. J. Lewis and R. G. Wilkins), p. 78, Interscience, 1960.

N. Sutin, The Kinetics of Inorganic Reactions in Solution, in *Annual Review of Physical Chemistry*, 1966, **17**, 119.

F. Basolo and R. G. Pearson, The Trans-Effect in Metal Complexes, in *Progress in Inorganic Chemistry* (ed. F. A. Cotton), Vol. 4, p. 381, Wiley, 1962.

F. Basolo and R. G. Pearson, *Mechanisms of Inorganic Reactions*, Second Edition, Wiley, 1967.

3. Oxidation–reduction reactions of metal ions

Electron transfer reactions involving metal ions and their complexes are of two basic types. These have become generally known as 'outer-sphere' and 'inner-sphere'. In outer-sphere processes the coordination shells of the metal ions remain intact during electron transfer. In inner-sphere processes electron transfer takes place through a bridging-group common to the coordination shells of both metal ions. Electron exchanges between substitution-inert reactants proceed via outer-sphere processes; electron transfers between reactants which are substitution-labile proceed by inner- or outer-sphere processes. One other possibility for electron transfer should be noted. Rather than being transferred directly, the electron may be released by the reductant to the solvent and subsequently transferred from the solvent to the oxidant, thus:

$$Cr^{2+} \rightarrow Cr^{3+} + e^-$$

$$Fe^{3+} + e^- \rightarrow Fe^{2+}$$

While this may take place in liquid ammonia with strong reducing agents, there is no evidence to show that such a mechanism is possible in aqueous solution.

Most transition elements, by having stable states differing by one in oxidation number, react with one another by one-equivalent steps. Differences of two between the stable oxidation states occur for some of the post-transition elements (e.g., Tl(I)–Tl(III), Sn(II)–Sn(IV), $[Hg(I)]_2$–2Hg(II), Pb(II)–Pb(IV)) and actinides (e.g., U(IV)–U(VI), Pu(IV)–Pu(VI)). Mechanistically, the main point at issue in such systems is whether their redox reactions occur through a single two-equivalent stage or through consecutive one-equivalent changes. The same interest is attached to reactions of Cr(VI) and Cr(III) where oxidation–reduction represents an overall transfer of three electrons.

For the purpose of discussion it is convenient to begin with an account of inner- and outer-sphere reactions. To follow this a limited but representative selection of specific reactions are considered in some detail* under the broad headings of one-equivalent–one-equivalent (e.g., (1)), two-equivalent–one-equivalent (e.g., (2)), two-equivalent–two-equivalent (e.g., (3)), and multi-equivalent (e.g., (4)) processes

$$\text{Co(III)} + \text{Ce(III)} \rightarrow \text{Co(II)} + \text{Ce(IV)} \tag{1}$$

$$\text{Tl(III)} + 2\text{Fe(II)} \rightarrow \text{Tl(I)} + 2\text{Fe(III)} \tag{2}$$

$$\text{U(IV)} + \text{Tl(III)} \rightarrow \text{U(VI)} + \text{Tl(I)} \tag{3}$$

$$\text{Cr(VI)} + 3\text{Np(V)} \rightarrow \text{Cr(III)} + 3\text{Np(VI)} \tag{4}$$

The related subject of metal-ion-catalysed reactions is then treated along with brief accounts of the reactions of metal ions with molecular oxygen, hydrogen, and carbon monoxide. Finally the interaction of metal ions and the hydrated electron is discussed.

Inner- and outer-sphere reactions

Outer-sphere reactions

In principle an outer-sphere reaction can be recognized if (*a*) the rate law is of the first order in both reactants, and the activated complex composes the intact coordination shells of both metal ions, (*b*) the coordination shell of either metal ion is sufficiently inert to substitution so that the rate of electron transfer is faster than the rate of substitution. Many electron transfer reactions are now regarded as of the outer-sphere type. Unfortunately the high lability of simple aquo ions normally precludes an assignment of mechanism on the grounds of either criterion (*a*) or (*b*).

The exchange of $[Fe(CN)_6]^{3-}$ and $[Fe(CN)_6{}^{4-}]$ affords a typical example of a process which occurs by direct electron transfer through

* Emphasis throughout is on redox reactions between *different* metal ions. Although mentioned, no separate account is given of isotopic exchange reactions, e.g.,

$$Co^{3+} + Co^{2+} \rightarrow Co^{2+} + Co^{3+}$$

$$[Fe(phen)_3]^{3+} + Fe^{2+} \rightarrow [Fe(phen)_3]^{2+} + Fe^{3+}$$

These are dealt with exhaustively in an excellent review (1962) by Sutin (see bibliography).

an outer-sphere activated complex. The rate of exchange, studied by isotopic labelling, is very rapid ($k \sim 10^3$ M^{-1} s^{-1} at 4°).[1] Ferrocyanide (a low-spin d^6 system) and ferricyanide (a low-spin d^5 system) are both inert to substitution. Application of the Franck–Condon principle has some interesting repercussions. The principle (which is fundamental to the interpretation of electronic spectra) states that electronic transitions are virtually instantaneous in comparison with atomic rearrangements. The normal length of the Fe—C bond in $[Fe(CN)_6]^{3-}$ is less than that in $[Fe(CN)_6]^{4-}$. When an electron is transferred from $[Fe(CN)_6]^{4-}$ to $[Fe(CN)_6]^{3-}$

$$[Fe(CN)_6]^{4-} + [Fe(CN)_6]^{3-} \rightarrow [Fe(CN)_6]^{3-} + [Fe(CN)_6]^{4-}$$

there is no change in the configuration of the atoms, Fe, C or N. Consequently the bond length of Fe—C in the newly-formed $[Fe(CN)_6]^{3-}$ ion will be longer than the equilibrium value and the length of the Fe—C bond in the product $[Fe(CN)_6]^{4-}$ ion will be shorter than the equilibrium value. In other words two vibrationally-excited ions will have been produced. Violation of the principle of conservation of energy is implied unless a prior rearrangement of the coordination shells of each ion takes place. In the formation of the activated complex, activation energy must be supplied to equalize the length of Fe—C bonds in the two ions. Since the bond lengths are not very dissimilar, little free energy of activation is required and exchange takes place rapidly. The slowness of the exchange between $[Co(NH_3)_6]^{2+}$ and $[Co(NH_3)_6]^{3+}$ ($k \leqslant 10^{-8}$ M^{-1} s^{-1} at 64°[2]) can hardly arise from the slight difference in Co—N bond lengths. However, these Co(III) complexes display widely different electronic configurations ($[Co(NH_3)_6]^{2+}$ is $t_{2g}{}^5 e_g{}^2$ whereas $[Co(NH_3)_6]^{3+}$ is $t_{2g}{}^6$), inferring that changes in the electronic arrangement, in addition to coordination rearrangements, must needs occur before electron transfer can take place. The argument can be extended to systems in which there is a net chemical change and the standard free energy of reaction is negative. For such cases reorientation of the coordination shells of the reactants is less stringent since the vibrational excitation energy of the products is liberated as part of the standard free energy of reaction. Stated differently, as the reaction becomes more exothermic (i.e., $\Delta G°$ becomes more negative) the structure of the transition state will more closely resemble that of the reactants, the free energy of activation will decrease and the rate of reaction will increase. Certainly, oxidation–reduction reactions between ions of

Table 3.1

Comparison of observed and calculated rate constants for some outer-sphere electron transfer reactions at 25°*

	k_{11}, $M^{-1}\ s^{-1}$	k_{22}, $M^{-1}\ s^{-1}$	K_{12}	k_{12} obs., $M^{-1}\ s^{-1}$	k_{12} calc.,† $M^{-1}\ s^{-1}$
Fe^{3+}–Cr^{2+}	4·0	$\leqslant 2 \times 10^{-5}$	2×10^{19}	$2{\cdot}3 \times 10^{3}$	$\leqslant 6 \times 10^{5}$
Ce(IV)-$[W(CN)_8]^{4-}$	4·4	7×10^{4}	$1{\cdot}6 \times 10^{15}$	$>10^{8}$	$6{\cdot}1 \times 10^{8}$
Ce(IV)-$[Fe(CN)_6]^{4-}$	4·4	3×10^{2}	$6{\cdot}3 \times 10^{12}$	$1{\cdot}9 \times 10^{6}$	$6{\cdot}0 \times 10^{6}$
Ce(IV)-$[Mo(CN)_8]^{4-}$	4·4	3×10^{4}	$6{\cdot}3 \times 10^{10}$	$1{\cdot}4 \times 10^{7}$	$1{\cdot}3 \times 10^{7}$
$[IrCl_6]^{2-}$-$[W(CN)_8]^{4-}$	2×10^{5}	7×10^{4}	$3{\cdot}9 \times 10^{6}$	$6{\cdot}1 \times 10^{7}$	$8{\cdot}1 \times 10^{7}$
$[IrCl_6]^{2-}$-$[Fe(CN)_6]^{4-}$	2×10^{5}	3×10^{2}	$1{\cdot}7 \times 10^{4}$	$3{\cdot}8 \times 10^{5}$	$5{\cdot}7 \times 10^{5}$
$[IrCl_6]^{2-}$-$[Mo(CN)_8]^{4-}$	2×10^{5}	3×10^{4}	$1{\cdot}6 \times 10^{2}$	$1{\cdot}9 \times 10^{6}$	$1{\cdot}0 \times 10^{6}$
$[Mo(CN)_8]^{3-}$-$[W(CN)_8]^{4-}$	3×10^{4}	7×10^{4}	$2{\cdot}5 \times 10^{4}$	$5{\cdot}0 \times 10^{6}$	$1{\cdot}7 \times 10^{7}$
$[Mo(CN)_8]^{3-}$-$[Fe(CN)_6]^{4-}$	3×10^{4}	3×10^{2}	$1{\cdot}1 \times 10^{2}$	$3{\cdot}0 \times 10^{4}$	$2{\cdot}7 \times 10^{4}$
$[Fe(CN)_6]^{3-}$-$[W(CN)_8]^{4-}$	3×10^{2}	7×10^{4}	$2{\cdot}4 \times 10^{2}$	$4{\cdot}3 \times 10^{4}$	$5{\cdot}1 \times 10^{4}$
$[Co(NH_3)_6]^{3+}$–V^{2+}	$<3{\cdot}3 \times 10^{-12}$	1×10^{-2}	$1{\cdot}3 \times 10^{6}$	$3{\cdot}3 \times 10^{-3}$	$<1{\cdot}5 \times 10^{-4}$
$[Co(NH_3)_6]^{3+}$-$[Ru(NH_3)_6]^{2+}$	$<3{\cdot}3 \times 10^{-12}$	3	$1{\cdot}4 \times 10^{-2}$	1×10^{-2}	$<3 \times 10^{-7}$
Co^{3+}–Fe^{2+}	5·0	4·0	10^{20}	$4{\cdot}2 \times 10^{1}$	6×10^{6}

* From N. Sutin, *Proceedings of Symposium on Exchange Reactions*, Brookhaven National Laboratory, International Atomic Energy Agency, Vienna, 1965, and references cited therein.

† Calculated from $k_{12} = (k_{11}k_{22}K_{12}f)^{1/2}$.

different metals (for which $\Delta G^\circ < 0$) are faster, on the whole, than the isotopic exchange reactions of either reactant. If the interaction of the orbitals involved in the electron transfer process is large enough for electron transfer to occur without hindrance but small enough to be neglected in calculating the free energy of activation, then a simple relationship applies between the rate constants of related outer-sphere processes. From the treatment of Marcus,[3] if k_{12} and K_{12} are the rate and equilibrium constant, respectively, for the electron transfer process then an interrelationship exists with the rate constants (k_{11} and k_{22}) for the exchange reactions of the reactants. For example, if k_{12} and K_{12} refer to the reaction

$$[Fe(CN)_6]^{4-} + [Mo(CN)_8]^{3-} \rightleftharpoons [Fe(CN)_6]^{3-} + [Mo(CN)_8]^{4-}$$

and k_{11} and k_{22} refer to the exchange reactions

$$[Fe(CN)_6]^{4-} + [Fe(CN)_6]^{3-} \rightleftharpoons [Fe(CN)_6]^{3-} + [Fe(CN)_6]^{4-}$$

and

$$[Mo(CN)_8]^{3-} + [Mo(CN)_8]^{4-} \rightleftharpoons [Mo(CN)_8]^{4-} + [Mo(CN)_8]^{3-}$$

then

$$k_{12} = (k_{11} k_{22} K_{12} f)^{1/2} \tag{3.1}$$

where

$$\log f = \frac{(\log K_{12})^2}{4 \log (k_{11} k_{22}/Z^2)}$$

and Z is a measure of the number of collisions occurring between two neutral species in unit volume of solution in unit time at the mean separation distance in the activated complex.

Table 3.1 shows a comparison between the rate constants observed and those calculated on the basis of eq. (3.1). With the exception of the systems $Co^{3+} + Fe^{2+}$ and $[Co(NH_3)_6]^{3+} + [Ru(NH_3)_6]^{2+}$, there is a remarkable agreement between the experimental results and the Marcus theory.

Inner-sphere reactions

In inner-sphere reactions substitution of the coordination shell of one of the metal ions occurs prior to electron transfer. For example, in the reaction

$$Cr_{aq}{}^{2+} + [XCo(NH_3)_5]^{2+} + 5H^+ \rightarrow CrX_{aq}{}^{2+} + Co_{aq}{}^{2+} + 5NH_4{}^+$$

the bridging ligand X is found in the coordination shell of the newly created Cr(III) ion.[4] When X is water or hydroxyl, labelled oxygen from the oxidant is found in the Cr(III) product. As the cobalt(III) complex and the Cr(III) product are inert to substitution, the electron transfer must proceed through a bridged intermediate formed as a result of substitution of X for water in the coordination shell of Cr(II):

$$[(NH_3)_5Co^{III}X]^{2+} + [Cr^{II}(H_2O)_6]^{2+} \rightarrow [(NH_3)_5Co^{III}{-}X{-}Cr^{II}(H_2O)_5]^{4+} + H_2O$$

Further support comes from the observation that, on the addition of radioactive Cl^- to the solution, no activity is found in the $CrCl_{aq}{}^{2+}$ product.*

Only in certain favourable cases can an inner-sphere mechanism be demonstrated unambiguously. Recently much attention has been focused on this problem. In the reduction of $[Fe(CN)_6]^{3-}$ with $[Co(CN)_5]^{3-}$ a product having the structure $[(NC)_5Co(NC)Fe(CN)_5]^{6-}$ has been isolated.[5] Other binuclear species have been detected as intermediates in oxidation–reduction reactions of metal aquo-ions. When pale blue solutions of V(IV) and Cr(II) are mixed a bright green colour develops instantly, then slowly fades to the blue purple colour characteristic of V^{3+} and Cr^{3+} in perchloric acid. The rates of formation and dissociation of the intermediate have been measured successfully using a flow apparatus.[6] Similarly in the Np(VI)+Cr(II) system a stable binuclear species, of composition $[O{-}Np^{V}{-}O{-}Cr^{III}(H_2O)_5]^{4+}$, has been separated by ion exchange and characterized directly by analysis.[7] In addition, kinetic evidence has revealed the existence of oxygen- or hydroxide-bridged dimers in the reactions between U(VI) and Cr(II),[8] Pu(VI) and Fe(II),[9] and V(III) and Cr(II).[10] Dimers are formed also from the interaction of Cr(VI) and Cr(II),[11] V(IV) and V(II),[12] and Fe(IV) and Fe(II).[13]

One system, the $[Co(CN)_5]^{3-}$-catalysed substitution of cyanide in various pentamminecobalt(III) complexes ($Co(NH_3)_5X$), has been studied in which inner- and outer-sphere mechanisms occur together.[14]

* A similar result to this has been recorded for the $CrX^{2+} + Cr^{2+}$ exchange. In this case if the Cr^{2+} is labelled initially all the activity is found in the CrX^{2+} product, and thus the exchange must proceed through an inner-sphere activated complex in which X is attached to both Cr atoms (H. Taube and E. L. King, *J. Amer. Chem. Soc.*, 1954, **76**, 4053). Since the net effect is the transfer of X, such processes are sometimes referred to as *atom transfer* reactions.

Two distinct patterns of behaviour were noted. When X (the ligand present in the complex) is Cl^-, N_3^-, NCS^- or OH^-, the stoichiometry of the reaction conforms to

$$Co(NH_3)_5X + 5CN^- \rightarrow Co(CN)_5X + 5NH_3$$

and the rate law is

$$\text{rate} = k_i[Co(NH_3)_5X][Co(CN)_5{}^{3-}]$$

However, when $X = PO_4{}^{3-}$, $CO_3{}^{2-}$, $SO_4{}^{2-}$, NH_3, OAc^- or other carboxylates, a different stoichiometry and rate law obtains

$$Co(NH_3)_5X + 6CN^- \rightarrow [Co(CN)_6]^{3-} + 5NH_3 + X$$

$$\text{rate} = k_0'[Co(NH_3)_5X][Co(CN)_5{}^{3-}][CN^-]$$

In the first case it seems clear that the substitution takes place by an inner-sphere electron transfer between $[Co^{II}(CN)_5]^{3-}$ and $Co^{III}(NH_3)_5X$ through the bridged intermediate

$$[(CN)_5Co^{II}—X—Co^{III}(NH_3)_5].$$

In the second case an outer-sphere mechanism operates, electron transfer occurring between $[Co^{II}(CN)_6]^{4-}$ (assumed to exist in equilibrium with $[Co^{II}(CN)_5]^{3-}$) and $Co^{III}(NH_3)_5X$

$$[Co^{II}(CN)_5]^{3-} + Co^{III}(NH_3)_5X \xrightarrow{k_i} [(CN)_5Co^{II}—X—Co^{III}(NH_3)_5] \rightarrow Co^{III}(CN)_5X$$

$$+CN^- \downarrow\uparrow K$$

$$[Co^{II}(CN)_6]^{4-} + Co^{III}(NH_3)_5X \xrightarrow{k_o} [(CN)_5Co^{II}(CN)(X)Co^{III}(NH_3)_5] \rightarrow [Co^{III}(CN)_6]^{3-}$$

The observed rate constant for the outer-sphere route is thus identified as $k_o' = k_oK$. Inspection of the kinetic data reveals the outer-sphere process to be largely independent of the nature of the ligand X. However, a gradual trend is apparent for k_o' to increase as the positive charge of the complex increases (5×10^2 M^{-2} s^{-1} for $[Co(NH_3)_5PO_4]$ to 8×10^4 M^{-2} s^{-1} for $[Co(NH_3)_6]^{3+}$). Inability to detect a significant contribution from the outer-sphere path is ascribed to the intervention of an inner-sphere path, whose rate constant, k_i, shows an expected sensitivity to the nature of X.

A transient intermediate of an inner-sphere path has been successfully detected by using as an oxidant $[Co(NH_3)_5NO_2]^{2+}$, a complex containing a N-bonded nitrito group.[15] The final product corresponds to the N-bonded $[Co(CN)_5NO_2]^{3-}$. Stopped-flow oscillograms reveal the presence of a transient intermediate which decays more slowly (half-life 7 s) than $[Co(NH_3)_5NO_2]^{2+}$ (half-life 0·1 s). The intermediate by exhibiting strong absorption at 380 mμ, a wavelength where the other species in the reaction are transparent, would seem to contain a O-bonded ligand (cf. $[Co(CN)_5OH_2]^{2-}$ which absorbs also at 380 mμ). Thus the sequence of steps is likely to be the attack of $[Co(CN)_5]^{3-}$ on an oxygen atom of the NO_2 group, followed by electron transfer via the intermediate $[(NH_3)_5Co—NO_2—Co(CN)_5]^-$ to generate $[Co(CN)_5ONO]^{3-}$ which then isomerizes to the stable form $[Co(CN)_5NO_2]^{3-}$. Similarly, direct evidence has been culled for a transient intermediate $[(NH_3)_5Co—CN—Co(CN)_5]^-$ produced by the attack of $[Co(CN)_5]^{3-}$ on the nitrogen atom of the C-bonded oxidant $[Co(NH_3)_5CN]^{2+}$. In this case the intermediate decays ultimately to $[Co(CN)_6]^{3-}$ through the unstable species $[Co(CN)_5NC]^{3-}$.

There is less clearly-defined evidence to suggest that the reaction of $[Co(NH_3)_5CN]^{2+}$ with Cr^{2+} proceeds by the intervention of the unstable species $[Cr(H_2O)_5NC]^{2+}$.[16] Also the complex $[Cr(H_2O)_5SCN]^{2+}$ has been characterized as the initial product of reaction between Cr^{2+} and Fe(III) in the presence of thiocyanate ions. This S-bonded thiocyanato complex is *not* formed when a solution of chromium(II) and thiocyanate is oxidized by Fe(III). In the presence of excess Cr(II) ions $[Cr(H_2O)_5SCN]^{2+}$ reverts to the stable N-bonded isomer $[Cr(H_2O)_5NCS]^{2+}$.[17] The same S-bonded complex is formed in the oxidation of chromium(II) with *cis*- and *trans*-$[Co(en)_2(OH_2)(NCS)]^{2+}$. The preparation of $[Cr(H_2O)_5SCN]^{2+}$ can be effected simply by the gradual addition of a 5×10^{-3} M Cr^{2+} solution to an equal volume of a well-stirred solution containing $5{\cdot}5 \times 10^{-3}$ M Fe(III) and $4{\cdot}5 \times 10^{-3}$ M SCN^-. The resulting solution is green whereas a solution of $[Cr(H_2O)_5NCS]^{2+}$ is purple.[18]

The reductions of $[Co(C_2O_4)_3]^{3-}$, $[Co(en)_2(OH_2)Cl]^{2+}$ and $[Co(NH_3)_3(OH_2)_2N_3]^{2+}$ by Fe(II) (like those of chromium(II)) proceed by inner-sphere mechanisms, giving rise to the detectable Fe(III) species, $FeC_2O_4{}^+$, $FeCl^{2+}$ and $FeN_3{}^{2+}$, respectively.[19]

In one investigation an inner-sphere mechanism has been established for a system in which both reactants and both products are substitution-labile. The rate of the chloride-catalysed oxidation of iron(II) by cobalt(III) has been studied in perchloric acid using a flow-technique.[20] On mixing a solution of Co(III), Co(II), and chloride ion with a solution of Fe(II), reaction was complete within a few milliseconds at the concentrations employed. $FeCl^{2+}$ was identified spectrophotometrically (at 336 mμ) as the immediate product of reaction. No change in absorbance was detected at 336 mμ, however, when a solution of Co(III) and Co(II) was mixed with one containing Fe(II) and chloride. These results affirm that $FeCl^{2+}$ is produced as an intermediate in the reaction between Fe^{2+} and $CoCl^{2+}$

$$CoCl^{2+} + Fe^{2+} \rightarrow FeCl^{2+} + Co^{2+}$$

and that this reaction occurs through an inner-sphere activated complex* having the structure $[Co{-}Cl{-}Fe_{aq}{}^{4+}]^{\ddagger}$.

If direct evidence is not forthcoming, assignment of mechanism can often be made on the basis of indirect criteria. The formation of a bridged activated complex by the replacement of a water molecule from the coordination shell of a hydrated metal ion should be attended by an increase in the entropy of activation. This effect, superimposed upon the normal decrease in entropy expected on forming an activated complex between charged reactants, requires the activation entropy of inner-sphere reactions to be less negative than that of outer-sphere reactions. Inner-sphere reactions should also be recognized by the considerable effect brought about by variations in the nature of the bridging group.[21] Thus the rates of Cr^{2+} reactions are more susceptible to variation in the nature of the oxidant than are the reactions of $[Cr(bipy)_3]^{2+}$. Similarly, reactions of Eu^{2+} are designated as inner-sphere whereas those of V^{2+} are outer-sphere.[22]† Also, azide ion is predicted to be much more efficient than

* A similar inner-sphere path exists in the chloride-catalysed $Fe^{3+} + Fe^{2+}$ exchange, viz., $FeCl^{2+} + Fe^{2+} \rightarrow Fe^{2+} + FeCl^{2+}$. Reaction via this route is only about three times faster than by the (outer-sphere) electron-transfer path $FeCl^{2+} + Fe^{2+} \rightarrow FeCl^{+} + Fe^{3+}$ (R. J. Campion, T. J. Conocchioli, and N. Sutin, *J. Amer. Chem. Soc.*, 1964, **86**, 4591).

† The rate constants for the oxidation of V^{2+} by Fe^{3+}, $FeOH^{2+}$, $FeCl^{2+}$, $FeNCS^{2+}$, and $FeN_3{}^{2+}$ are $1{\cdot}8 \times 10^4$, $\leqslant 4 \times 10^5$, $4{\cdot}6 \times 10^5$, $6{\cdot}6 \times 10^5$, and $5{\cdot}2 \times 10^5$ M^{-1} s^{-1}, respectively, at 25° and $\mu = 1$ M. The reaction between V^{2+} and Fe^{3+} is almost certainly of the outer-sphere type since the rate of water substitution in V^{2+} is much slower than the rate of electron transfer. However, while the unusually small effect of hydroxide and the similar effects of Cl^-, NCS^-, and $N_3{}^-$ would seem also to indicate an outer-sphere process, the suggestion has been made that these $FeX^{2+} + V^{2+}$ reactions proceed via inner-sphere, water-bridged activated complexes formed by the loss of a water molecule in $[Fe(H_2O)_6]^{3+}$ (B. R. Baker, M. Orhanovic, and N. Sutin, *J. Amer. Chem. Soc.*, 1967, **89**, 722).

thiocyanate as a bridging group in inner-sphere reactions.[23] For outer-sphere reactions there should be little or no distinction between azide and thiocyanate.* Inspection of the kinetic results available shows that the oxidations of Cr^{2+}, and probably Eu^{2+} and Fe^{2+}, by $[Co(NH_3)_5X]^{2+}$ (where $X = N_3^-$ or NCS^-) are of the inner-sphere type, with $k_{N_3^-}/k_{NCS^-}$ at 25° given by $\sim 2 \times 10^4$, $\sim 3 \times 10^2$, and $\geqslant 3 \times 10^3$, respectively. Oxidation of V^{2+} and $[Cr(bipy)_3]^{2+}$ by $[Co(NH_3)_5X]^{2+}$ are, on the other hand, outer-sphere reactions with $k_{N_3^-}/k_{NCS^-}$ at 25° equal to 43 and 4, respectively.

Inorganic bridging ligands in inner-sphere reactions

When Cr^{2+}, V^{2+}, $[Co(CN)_5]^{3-}$, $[Ru(NH_3)_6]^{2+}$, and $[Cr(bipy)_3]^{2+}$ are the reducing agents for $[Co(NH_3)_5X]^{2+}$ complexes then the order of rates, and thus bridging efficiency, is given by $X = I^- > Br^- > Cl^- > F^-$. Such an order is to be expected on considering the relative polarizabilities of the halide groups. On the other hand, an inversion of this order occurs with the reducing agents Fe^{2+} and Eu^{2+}. Also the reduction of $[Cr(NH_3)_5X]^{2+}$ and $[Cr(H_2O)_5X]^{2+}$ by Cr^{2+} conforms to a normal order whereas the reduction of $[Fe(H_2O)_5X]^{2+}$ by Fe^{2+} displays inversion. Thus the order of bridging efficiency depends principally on the reducing agent and, to a lesser extent, on the nature of the oxidizing agent. Factors other than ligand polarizability play a part in deciding this order. There is evidence to show that the rates of inner-sphere reactions, like those of the outer-sphere type, increase as the standard free energy change (ΔG°) becomes more negative.[17] Table 3.2 shows the good agreement obtained between the observed rate constants for a number of inner-sphere reactions (of $[Co(NH_3)_5X]^{2+}$ and $[Co(en)_2(NCS)Cl]^+$) and those calculated on the basis of the Marcus theory of outer-sphere reactions (p. 95). It is interesting that the reduction of an excess of *cis*-$[Co(NH_3)_4(N_3)_2]^+$

* This comes about from the observation that the exchange of Cr^{2+} with CrN_3^{2+} ($k > 1{\cdot}2\ M^{-1}\ s^{-1}$ at 0°) is very much faster than that between Cr^{2+} and $CrNCS^{2+}$ ($k = 1{\cdot}8 \times 10^{-4}\ M^{-1}\ s^{-1}$ at 27°). Both exchanges are of the inner-sphere type. The considerable difference in rate is probably due to differences in the structure of their respective activated complexes: the symmetrical $[(H_2O)_5Cr^{III}{-}N{=}N{=}N{-}Cr^{II}(H_2O)_5{}^{4+}]^{\ddagger}$ and the unsymmetrical $[(H_2O)_5Cr^{III}{-}N{=}C{-}S{-}Cr^{II}(H_2O)_5{}^{4+}]^{\ddagger}$. The decomposition of the latter complex yields Cr^{2+} and the unstable S-bonded isomer $CrSCN^{2+}$.[23]

Table 3.2

Rate constants* for some reactions of chromium(II) and iron(II) with cobalt(III) complexes, at 25° and $\mu = 1{\cdot}0$ M

Reaction	log k_{obs}	log k_{calc}†
$[Co(NH_3)_5Cl]^{2+} + Fe^{2+}$	−2·9	. . .
$[Co(NH_3)_5Cl]^{2+} + Cr^{2+}$	6·4	6·6
$[Co(NH_3)_5F]^{2+} + Fe^{2+}$	−2·2	. . .
$[Co(NH_3)_5F]^{2+} + Cr^{2+}$	5·9	5·5
$[Co(NH_3)_5N_3]^{2+} + Fe^{2+}$	−2·0	. . .
$[Co(NH_3)_5N_3]^{2+} + Cr^{2+}$	5·5	5·7
cis-$[Co(en)_2(NCS)Cl]^+ + Fe^{2+}$	−3·8	. . .
cis-$[Co(en)_2(NCS)Cl]^+ + Cr^{2+}$	6·3	5·7
trans-$[Co(en)_2(NCS)Cl]^+ + Fe^{2+}$	−3·9	. . .
trans-$[Co(en)_2(NCS)Cl]^+ + Cr^{2+}$	6·4	5·6

* From ref. (17), and references cited therein.

† Calculated from $k_{12}/k_{13} = (k_{22}K_{12}/k_{33}K_{13})^{1/2}$ (obtained by assuming that the f terms cancel).

(or *cis*-$[Co(en)_2(N_3)_2]^+$) by chromium(II) takes place by two parallel routes involving single- and double-bridged activated complexes[24]

$$Cr^{2+} + cis\text{-}[Co(NH_3)_4(N_3)_2]^+ \begin{cases} \nearrow cis\text{-}Cr(N_3)_2^+ \\ \searrow CrN_3^{2+} \end{cases}$$

The standard free energy changes favour the inverted order in both the Fe^{2+} and Cr^{2+} cases. Apparently, trends in ligand polarizability and $\Delta G°$ exert opposing effects with the result that the bridging order is determined by polarizability trends in Cr^{2+} reductions but by free energy considerations in Fe^{2+} reductions. It should be noted, however, that a normal order of bridging is observed in the exchange of Cr^{2+} and CrX^{2+} but an inverted order for the $Fe^{2+} + FeX^{2+}$ exchange. Since both exchanges are of the inner-sphere type, for which $\Delta G° = 0$, the reversal of order expected on grounds of ligand polarizabilities in the case of the iron exchange is attributed to π-bonding factors.

The relative efficiencies of oxoanions as bridging groups has been investigated in detail by examining the rates of reaction of Cr(II),

V(II), Eu(II), and Ti(III) with oxoanion complexes of pentaammine- and tetraamminecobalt(III).[25] The complexes employed include metaborato, carbonato, nitro, nitrito, nitrato, sulphito, sulphato, aquosulphato, thiosulphato, selinito, selanato, and phosphato. A relationship was observed between the rate of reduction of the complex and the position in the periodic table of the central atom of the oxoanion ligand: the rate increases to a maximum from Group III to V and then decreases. Also, for a given group, the rate increases as the atomic weight of the central atom increases.

Organic bridging groups in inner-sphere reactions

An understanding of the role of organic ligands as bridging groups in electron transfer reactions has come about as a result of a very extensive examination of the rates of chromium(II) (and other reducing agents) with over a hundred carboxylatopentaamminecobalt(III) complexes.[26] Analysis of the results shows that the organic ligand functions as a mediator for electron transfer if it contains a conjugated bond system, or if it contains groups which are capable of associating strongly with the reducing agent. Furthermore, three classes of reaction

Table 3.3

Rate parameters for the reaction of Cr^{2+} with carboxylato-pentaamminecobalt(III) complexes (at 25° and $\mu = 1$ M)*

Ligand	k, M^{-1} s^{-1}	$\Delta H^{\ddagger}$, kcal mole^{-1}	$\Delta S^{\ddagger}$, cal deg^{-1} mole^{-1}
CH_3COO	0·18	3·0	−52
$ClCH_2COO$	0·10	7·9	−37
Cl_2CHCOO	0·074	2·5	−55
F_3CCOO	0·052	...	...
Benzoato	0·14	4·9	−46
o-chlorobenzoato	0·074	6·0	−43
p-chlorobenzoato	0·21	10·0	−28
p-hydroxybenzoato	0·13	9·6	−30
Isophthalato	0·13	2·1	−56
HCOO	7	...	...

* From H. Taube, in *Mechanisms of Inorganic Reactions*, Advances in Chemistry Series (ed. R. F. Gould), No. 49, p. 107, American Chemical Society, 1965, and references cited therein.

of Cr(II) with $[Co(NH_3)_5CO_2R]^{2+}$ complexes can be discerned: these have been referred to as *adjacent attack*, *adjacent attack with chelation*, and *remote attack*.

Table 3.3 contains rate parameters for those ligands which have no propensity for chelation with Cr(II) and which function as electron mediators only by their ability to bridge directly across the oxidizing and reducing agents, for example:

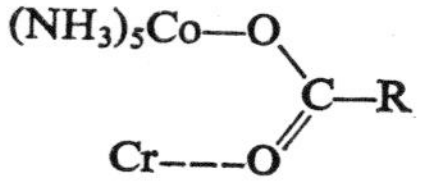

$$\begin{array}{l}(NH_3)_5Co\text{—}O \\ \qquad\qquad\quad \diagdown \\ \qquad\qquad\quad C\text{—}R \\ \qquad\qquad\quad /\!\!/ \\ Cr\text{---}O \end{array}$$

The electron transfer path in such cases is simply through the sequence of bonds Cr—O—C—O—Co. Although the ligands have different base strengths (as reflected by variations of 10^4 to 10^5 in the dissociation constants of their corresponding organic acids), their individual influences on the rate of oxidation of Cr(II) are slight, and it is evident that (with the exception of the formate group) the rates are remarkably similar, and within the range $k = 0{\cdot}05$ to $0{\cdot}2\ M^{-1}\ s^{-1}$ at 25°. The small variation in rate probably comes about because of steric effects; the marked influence of formate supports this interpretation.

The rate of reaction is enhanced if the ligand is able to chelate with the reducing agent. Examples of such chelating ligands are α-hydroxy acids, like glycolate (1); and those containing a carbonyl group (2) or a hydroxy group (3) *ortho* to the coordinated carboxyl group.

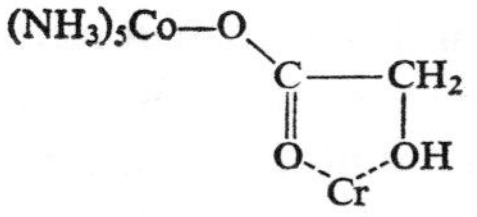

(1) $(NH_3)_5Co$—O—C(=O···Cr)—CH_2—OH···Cr

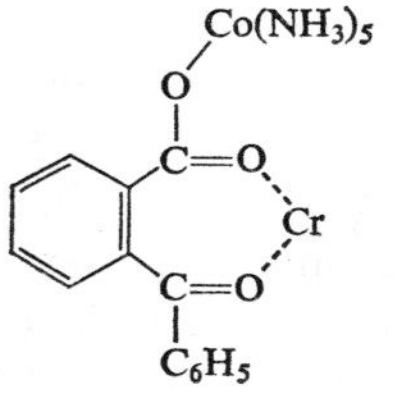

(2) $Co(NH_3)_5$—O—C(=O···Cr)—C_6H_4—C(=O···Cr)—C_6H_5

(3) $Co(NH_3)_5$—O—C(=O—Cr)—C_6H_4—O—Cr

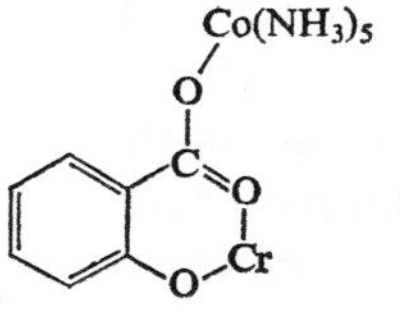

In these examples chromium forms part of a five-, seven- or six-membered ring system. The efficiency of such ligands in promoting electron transfer from chromium to cobalt is probably related, in the first place, to the production of a firm route along which an electron may be transferred.

Those reactions in which the reducing agent is believed to attack the bridging ligand at a position remote from the coordinated carboxyl group constitute the third class. For such systems the rate is much greater than for reactions involving 'normal' or chelating ligands and the rate law commonly contains a hydrogen-ion dependent term. In most instances a conjugated path exists between the coordinated carboxylate group and a remote functional group, as is typified by the fumarato (4) and 4-carboxylatopyridine complexes (5)

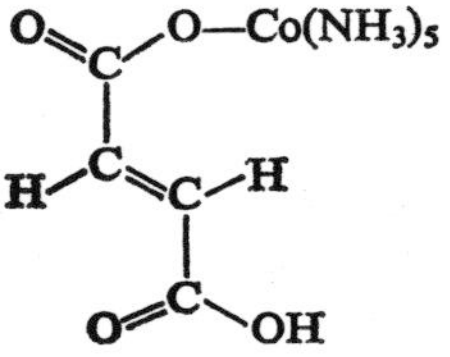

$O{=}C(-O{-}Co(NH_3)_5){-}CH{=}CH{-}C({=}O){-}OH$

(4)

$O{=}C(-O{-}Co(NH_3)_5){-}C_5H_4N$

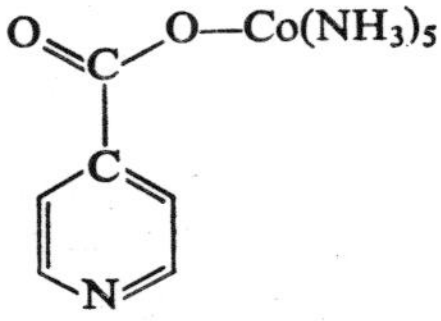

(5)

When electron transfer occurs from the reducing agent to the functional group of the organic ligand there is evidence to show that a radical-ion intermediate (conjugatively-stabilized) is formed. Remote attack is impossible unless the ligand is reducible.[27]

One-equivalent–one-equivalent reactions

The majority of redox reactions between metal ions are of the one-equivalent type. These simple reactions serve as models for more complicated systems and their study has proved invaluable in developing an understanding of the process of electron transfer in solution. Of particular importance has been the knowledge gained concerning the influence of anions: indeed, some of the more intimate details of this aspect have been dealt with in the previous sections. At this stage it will suffice to describe the results of investigations on two typical systems, the oxidations of cerium(III) and iron(II) by cobalt(III).

The oxidation of cerium(III) by cobalt(III) has received detailed attention.[28] Of particular note is the thorough investigation of the effect of simple anions on the rate of reaction in perchlorate media. Kinetic data for the reaction between Ce(III) and Co(III) perchlorates, in the absence of added anions, were obtained by following the disappearance of Co(III) at its absorption maximum of 605 mμ. At this wavelength there is no interference from Ce(IV) or Ce(III), and only a slight correction is needed for Co(II) absorption. The range of Ce(III) and Co(III) concentrations employed were $8{\cdot}2 \times 10^{-3}$ to $8{\cdot}2 \times 10^{-4}$ M and $5{\cdot}6 \times 10^{-4}$ to $2{\cdot}6 \times 10^{-4}$ M, respectively, at a constant ionic strength of 1·0 M and over a temperature range of 17·5 to 31·4°. Although the absorption spectrum of Ce(III) is uninfluenced by changes in hydrogen-ion concentration, temperature and ionic strength variations produce effects at the 296 mμ maximum. This is taken to indicate the formation of a complex between Ce(III) and perchlorate ion; the determination of the equilibrium constant and thermodynamic parameters for

$$Ce^{3+} + ClO_4^- \overset{K_2}{\rightleftharpoons} CeClO_4^{2+}$$

constituted the first stage of the investigation.

The reaction is stoichiometric

$$Co(III) + Ce(III) \rightarrow Co(II) + Ce(IV)$$

and kinetically of simple second order (Figs. 3.1 and 3.2); the addition of a large excess of Co(II) ($[Co(II)]_0/[Co(III)]_0 > 200$) does not influence the rate and thus any back-reaction is insignificant. The rate of reaction shows a marked inverse dependence on acidity over the range of perchloric acid concentrations 0·2 to 1·0 M at constant ClO_4^- concentration ($k' = a/[HClO_4]$ where k' is the observed second-order rate constant). The rate increases with increase in perchlorate ion concentration at constant acidity. To show this effect the ionic strength was maintained at 1·0 M and the ClO_4^- concentration varied by substitution of lanthanum perchlorate for sodium perchlorate. The results show a dependence of observed rate constant with ClO_4^- concentration given by $1/k' = b + c/[ClO_4^-]$ where b and c are constants. Since the possibility of a reacting hydrolysed species of Ce(III) is ruled out from previous studies, the hydrogen-ion dependence clearly demonstrates the existence of a hydrolysed form of Co(III)

$$Co^{3+} + H_2O \overset{K_1}{\rightleftharpoons} CoOH^{2+} + H^+$$

Obviously the optimum conditions for measurements of K_1 are low hydrogen-ion concentrations, but under these conditions the reduction of Co(III) by water becomes appreciable. However, by careful experimentation a value was obtained for K_1 of $1{\cdot}75 \times 10^{-2}$ M at 25°.

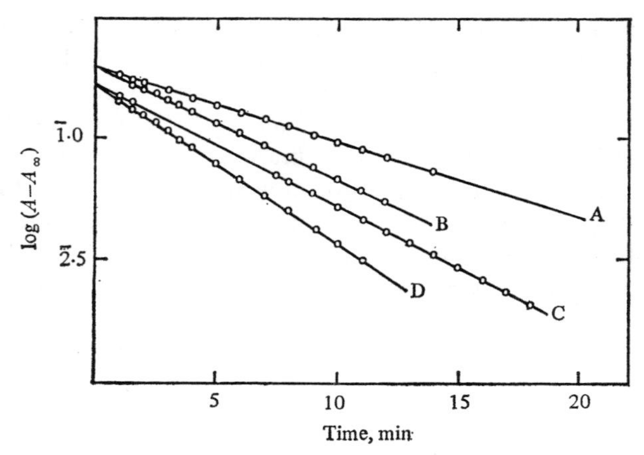

Fig. 3.1. Co(III) + Ce(III) reaction: first-order dependence of rate on Co(III) concentration at 17·5° ($[HClO_4]$ = 1·0 M, μ = 1·0 M). Ce(III) concentrations: A = 4·10 × 10^{-3} M, B = 5·73 × 10^{-3} M, C = 6·56 × 10^{-3} M, D = 8·19 × 10^{-3} M; cell length = 11·4 cm; λ = 605 mμ. A and A_∞ are, respectively, the absorbances at a given time and after complete reaction. From L. H. Sutcliffe and J. R. Weber, *Trans. Faraday Soc.*, 1956, **52**, 1225.

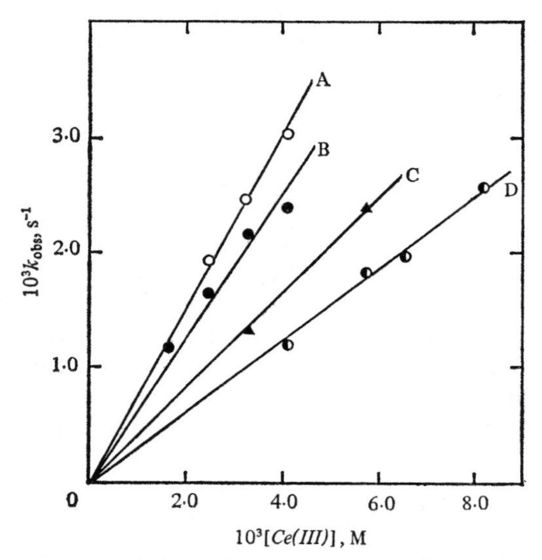

Fig. 3.2. Co(III) + Ce(III) reaction: first-order dependence of rate on Ce(III) concentrations at 17·5° (μ = 1·0 M). $HClO_4$ concentrations: A = 0·40 M, B = 0·50 M, C = 0·80 M, D = 1·00 M. k_{obs} corresponds to the slopes of plots such as Fig. 3.1. From L. H. Sutcliffe and J. R. Weber, *Trans. Faraday Soc.*, 1956, **52**, 1225.

The observed dependence of rate on the acid and perchlorate concentrations is consistent only with

$$CoOH^{2+} + CeClO_4^{2+} \xrightarrow{k} Co(II) + Ce(IV)$$

as the rate-controlling step. The participation of a perchlorate complex in a redox reaction is rather unusual. By assuming that the total Ce(III) and Co(III) concentrations are $[Ce^{3+}] + [CeClO_4^{2+}]$ and $[Co^{3+}] + [CoOH^{2+}]$, respectively, it can be deduced that

$$1/k' = [H^+]/kK_1 + [H^+]/kK_1 K_2[ClO_4^-]$$

if K_1 is less than $[H^+]$. Consequently for constant $[ClO_4^-]$, $a = kK_1K_2[ClO_4^-]/(1 + K_2[ClO_4^-])$, and for constant $[HClO_4]$, $b = [H^+]/kK_1$ and $c = [H^+]/kK_1 K_2$. These relationships enable k, the true rate constant, to be calculated as 95 M^{-1} s^{-1} at 25°. The corresponding enthalpy and entropy of activation are 19 kcal mole^{-1} and 14 cal mole^{-1} deg^{-1}, respectively.

A later spectrophotometric study of Co(III) perchlorate solutions[29] has shown that low acidities and high concentrations favour the formation of a complex Co(III) species whose ultraviolet absorption spectrum differs considerably from that of the Co^{3+} ion. It is suggested that the hydrolysed species first formed undergoes slow dimerization:

$$Co^{3+} + H_2O \rightleftharpoons CoOH^{2+} + H^+ \qquad \text{fast}$$

$$\left.\begin{array}{l} Co^{3+} + CoOH^{2+} \rightarrow Co{-}O{-}Co^{4+} + H^+ \\ CoOH^{2+} + CoOH^{2+} \rightarrow Co{-}O{-}CoOH^{3+} + H^+ \end{array}\right\} \text{slow}$$

The basic Co—O—Co structure is considered more likely than that postulated for the iron dimer, $Fe(OH)_2Fe^{4+}$. Further hydrolysis and/or polymerization is not excluded. To some extent these findings are supported by a kinetic examination of the reduction of Co(III) by water[30], in which reaction the rate is dependent on $[Co(III)]^{3/2}$ and inversely dependent on $[H^+]^2$. In this case the slow stage is thought to be the reaction of hydrolysed and dimeric species

$$CoOH^{2+} + HOCo{-}O{-}CoOH^{2+} \rightarrow 3Co^{2+} + 2OH^- + HO_2$$

The addition of fluoride, nitrate,[31] or sulphate[32] ions causes an increase in the rate of the Co(III) + Ce(III) reaction, the effect decreasing in the order $F^- > SO_4^{2-} > NO_3^-$. The reactive species are identified as CoF^{2+}; $CoSO_4^+$ (at 20°, $Co(SO_4)_2^-$ is the preponderant reactive complex at higher temperatures) and $CoNO_3^{2+}$, respectively. A recent and particularly interesting study of the formation of the monochloro complex of Co(III), using flow techniques, has given a considerably higher value for K_1 of 0·22 M at 25° (see p. 55).

The oxidation of iron(II) by cobalt(III) in perchloric acid solutions takes place at a rate which is just accessible to conventional spectrophotometry.[33] To obtain kinetic data, the reactant concentrations were made equal, and the reactant solution was sampled (typically at 30–50-second intervals) and run into a quenching solution made up of ammoniacal 2,2′-bipyridine. The absorbance (at 522 mμ) due to the tris-2,2′-bipyridyl complex of Fe(II) is proportional to the concentration of unreacted Fe(II). A typical run is shown as Fig. 3.3. Under the conditions employed the rate law

$$-\mathrm{d}[Fe(II)]/\mathrm{d}t = -\mathrm{d}[Co(III)]/\mathrm{d}t = k'[Co(III)][Fe(II)]$$

has the integrated form

$$1/[Fe(II)] = k't + 1/[Fe(II)]_0$$

where k' is related to the half-life of reaction by

$$k' = 1/[Fe(II)]_0 t_{1/2}$$

The accurate second-order nature of the reaction is preserved over a 60-fold range of concentration and there is no evidence for the participation of a polymerized species of cobalt(III). (Similarly there is no indication of such a species from the results of the Co(II) + Co(III) exchange.) The reaction displays a marked inverse dependence on hydrogen-ion concentration and plots of k' versus $1/[H^+]$ are linear. This observation suggests that the reaction involves a hydrolysed species of Co(III) and that the rate equation is better treated in terms of

$$-\mathrm{d}[Fe(II)]/\mathrm{d}t = k_0[Co^{3+}][Fe^{2+}] + k_1[CoOH^{2+}][Fe^{2+}]$$

where the observed rate constant k' is given by

$$k' = k_0 + k_1 K_1/[H^+]$$

and K_1 is the hydrolysis constant of Co^{3+}. The intercept and slope values of a plot of k' versus $1/[H^+]$ were used to obtain the individual rate constants, k_0 and k_1. At 0° and $\mu = 1$ M, k_0 and k_1 are 10 and 6500 M^{-1} s^{-1}, respectively. Thus it is clear that the greater part of reaction proceeds via $CoOH^{2+}$. This is in line with the Fe(II)+ Fe(III) and Co(II) + Co(III) exchanges, where the preferred paths are again shown to be those involving hydrolysed species. Sulphate ion has a significant effect on the reaction, due most likely to the introduction of the step

$$CoSO_4^+ + Fe^{2+} \xrightarrow{k_2} Co^{2+} + FeSO_4^+$$

Assuming this to be the case, k_2 has a value of 4900 $M^{-1}\ s^{-1}$ at 0°.

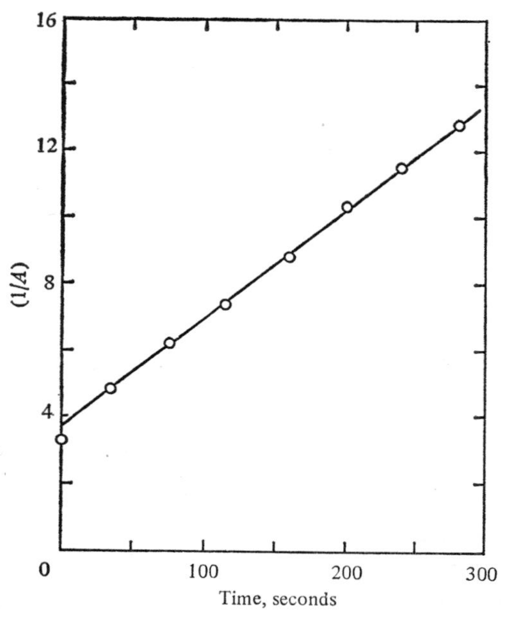

Fig. 3.3. Co(III) + Fe(II) reaction: second-order plot showing first-order dependence of rate on Co(III) and on Fe(II) concentrations at 0°; [*Fe(II)*] = [*Co(III)*] = 1·13 × 10^{-4} M, [$HClO_4$] = 0·285 M, μ = 1·0 M. The reciprocal of the absorbance of $[Fe(bipy)_3]^{2+}$ at 522 mμ is plotted versus time. From L. E. Bennett and J. C. Sheppard, *J. Phys. Chem.*, 1962, **66**, 1275.

The rate constants for the elementary steps of the Co(III) + Ce(III) and Co(III)+Fe(II) reactions are collected in Table 3.4 which includes also similar data for the Fe(III) + Cr(II) and Ce(IV) + Fe(II) systems. From these (and similar) results it can be seen what a marked influence anions can exert on the rate of redox reactions of metal aquo-ions (see, however, lower footnote on p. 99).

Table 3.4

Some electron-transfer reactions of transition metal ions

Reaction	*T*, °C	*k*, M^{-1} s^{-1}	Reaction	*T*, °C	*k*, M^{-1} s^{-1}
*Co(III)–Ce(III)**			*Fe(III)–Cr^{2+}§*		
$CoOH^{2+} + CeClO_4^{2+}$	25	95	$Fe^{3+} + Cr^{2+}$	25	$2{\cdot}3 \times 10^3$
$CoOH^{2+} + CeNO_3^{2+}$	25	93	$FeOH^{2+} + Cr^{2+}$	25	$3{\cdot}3 \times 10^6$
$CoOH^{2+} + CeF^{2+}$	25	$8{\cdot}5 \times 10^3$	$FeCl^{2+} + Cr^{2+}$	25	2×10^7
$CoSO_4^{+} + CeClO_4^{2+}$	25	$\leqslant 250$			
$CoSO_4^{+} + Ce^{3+}$	25	$\leqslant 350$	*Ce(IV) + Fe^{2+}‖*		
$Co^{3+} + CeSO_4^{+}$	25	$\leqslant 200$	$CeOH^{3+} + Fe^{2+}$	0	400
$CoSO_4^{+} + CeSO_4^{+}$	25	$\leqslant 2 \times 10^3$	$Ce(OH)_2^{2+} + Fe^{2+}$	0	$3{\cdot}23 \times 10^3$
			$CeSO_4^{2+} + Fe^{2+}$	0	5×10^3
Co(III)–Fe^{2+}†					
$Co^{3+} + Fe^{2+}$	0	10			
$CoOH^{2+} + Fe^{2+}$	0	$6{\cdot}5 \times 10^3$			
$CoSO_4^{+} + Fe^{2+}$	0	$4{\cdot}9 \times 10^3$			

* From ref. (31) and (32).
† From ref. (33).
§ From G. Dulz and N. Sutin, *J. Amer. Chem. Soc.*, 1964, **86**, 829.
‖ From M. G. Adamson, F. S. Dainton, and P. Glentworth, *Trans. Faraday Soc.*, 1965, **61**, 689.

Two-equivalent–one-equivalent reactions

Reactions of thallium(III) and thallium(I)

One-equivalent oxidants like Co(III), Mn(III), $[IrCl_6]^{2-}$ and $[Fe(phen)_3]^{3+}$ oxidize Fe^{2+} rapidly in a single step. By way of contrast the reactions of Fe^{2+} with two-equivalent oxidants are more complex and must involve unstable oxidation states of iron or the oxidant. The oxidation of Fe(II) by Tl(III) in aqueous perchloric acid (at ionic strengths between 3 and 6 M) is retarded by Fe(III).[34] Two successive one-electron transfers are indicated by the rate law

$$\frac{-\mathrm{d}[Fe(II)]}{\mathrm{d}t} = \frac{2k_1 k_3[Tl(III)][Fe(II)]^2}{k_2[Fe(III)] + k_3[Fe(II)]}$$

with both Fe(II) and Fe(III) competing for the Tl(II) intermediate

$$\mathrm{Tl(III) + Fe(II)} \underset{k_2}{\overset{k_1}{\rightleftharpoons}} \mathrm{Tl(II) + Fe(III)}$$

$$\mathrm{Tl(II) + Fe(II)} \xrightarrow{k_3} \mathrm{Tl(I) + Fe(III)}$$

The alternative two-electron transfer mechanism

$$\mathrm{Tl(III) + Fe(II) \rightleftharpoons Tl(I) + Fe(IV)}$$

$$\mathrm{Fe(IV) + Fe(II) \rightarrow 2Fe(III)}$$

involving Fe(IV), must be considered unimportant since the addition of Tl(I) has no effect on the rate. However, when the oxidant is chlorine, hypochlorous acid, hydrogen peroxide, or ozone there is evidence to suggest that Fe(IV) is produced as an intermediate. The method used in this investigation [13] was to mix the reactants in a flow apparatus and monitor the formation of Fe^{3+}, and the dissociation of any $FeCl^{2+}$ or dimeric $(FeOH)_2{}^{4+}$ produced in the reaction, at 240–280, 336, and 335 mμ, respectively. $FeCl^{2+}$ and $(FeOH)_2{}^{4+}$, while possessing similar absorption spectra, were distinguishable as a result of the dependence of their dissociation rates on acidity: the rate of dissociation of $(FeOH)_2{}^{4+}$ increases with increasing acidity whereas the dissociation of $FeCl^{2+}$ decreases with increasing acidity. The ion $FeOH^{2+}$ proved impossible to distinguish from Fe^{3+} because of its rapid conversion to Fe^{3+} by hydrogen ion. When the oxidant is HOCl or O_3 a substantial amount of dimeric $(FeOH)_2{}^{4+}$ results, suggesting that the following scheme obtains

$$\mathrm{Fe^{2+} + ox \rightarrow Fe(IV) + red}$$

$$\mathrm{Fe(IV) + Fe^{2+} \xrightarrow{fast} [Fe(III)]_2}$$

The rate of oxidation of V(IV) by Tl(III) is markedly reduced by the addition of V(V) but remains unaltered in the presence of Tl(I).[35] The rate law is given by

$$\frac{-\mathrm{d}[V(IV)]}{\mathrm{d}t} = \frac{k[Tl(III)][V(IV)]^2}{k'[V(V)] + [V(IV)]}$$

where $k \sim 1{\cdot}3$ M^{-1} min^{-1} and $k' \sim 42$ at 80°, $[H^+] = 1{\cdot}8$ M and $\mu = 3$ M. Thus a mechanism similar to the Tl(III)–Fe(II) one is demonstrated

$$\mathrm{Tl(III) + V(IV) \rightleftharpoons Tl(II) + V(V)}$$

$$\mathrm{Tl(II) + V(IV) \rightarrow Tl(I) + V(V)}$$

Establishment of this mechanism has a bearing on the Tl(I) + Tl(III) exchange reaction where, on the basis of the observed kinetics only (rate $= k[Tl(I)][Tl(III)]$), it is not possible to distinguish between a single-stage and a two-stage two-equivalent process. Since appreciable exchange takes place during the course of the reaction between Tl(III) and V(IV), it is concluded that Tl(II) is kinetically unimportant as an intermediate in the exchange. Otherwise the rate of reaction between Tl(III) and V(IV) would be altered.

Thallium(III) oxidizes V(III) so rapidly as to preclude a full kinetic investigation using conventional spectrophotometry. However, information on the mechanism has been obtained from measurements on the product solution.[36] No V(V) was detected as a product in perchloric acid, but up to 3 per cent of V(III) is converted to V(V) in sulphuric acid media in the presence of excess Tl(III). Moreover, this amount was shown to increase (to 10 per cent) when V(IV) was present initially. It is likely therefore that V(V) results from the conversion of V(IV) through Tl(II) as a reactive species:

$$\mathrm{Tl(II) + V(IV) \rightarrow Tl(I) + V(V)}$$

Accordingly the reaction is interpreted as proceeding by

$$\mathrm{Tl(III) + V(III) \rightarrow Tl(II) + V(IV)} \qquad \text{(one-equivalent change)}$$

$$\mathrm{Tl(II) + V(III) \rightarrow Tl(I) + V(IV)}$$

$$\mathrm{Tl(II) + V(IV) \rightarrow Tl(I) + V(V)}$$

$$\mathrm{V(V) + V(III) \rightarrow 2V(IV)}$$

rather than by

$$\mathrm{Tl(III) + V(III) \rightarrow Tl(I) + V(V)} \qquad \text{(two-equivalent change)}$$

$$\mathrm{V(V) + V(III) \rightarrow 2V(IV)}$$

It should be noted that changes in hydrolysis present a barrier to a single-stage two-equivalent step owing to the fact that V(IV) exists predominantly as VO^{2+} whilst VO_2^{+} is the stable form of V(V). Such restrictions do not arise in the case of U(V) and U(VI) since the relevant species are UO_2^{+} and UO_2^{2+}, respectively, and the oxidation of U(IV) by Tl(III) could be of the two-equivalent type. More recently, the development of fast reaction techniques has allowed a study of the kinetics of the reaction between V(III) and Tl(III) in acid perchlorate media at temperatures between 0·5 and 25°.[37] The course of the reaction was followed from the changes in absorbance at 760 mμ, the absorption peak for V(IV); 70–90 per cent completion of reaction corresponded to 25–30 s. The rate law is of simple first-order in both

Tl(III) and V(III). A determination of the rate constant for the V(V)+V(III) reaction reveals that this process is too slow (by a factor of ~14) to account for the observed rate of formation of V(IV). Also it was shown that the normally slow reaction between Tl(III) and Fe(II) can be induced by V(III). It follows that Fe(II) must react with an active intermediate and, since V(V) is ruled out from kinetic considerations, this species is identified as Tl(II).

A recent report[38] has shown that the oxidation of V(II) by Tl(III) takes place at a lower rate than the oxidation of V(III) and occurs predominantly through a two-equivalent change

$$\mathrm{V(II) + Tl(III) \rightarrow V(IV) + Tl(I)}$$

with contributions also from

$$\mathrm{V(II) + Tl(III) \rightarrow V(III) + Tl(II)}$$

$$\mathrm{V(III) + Tl(III) \rightarrow V(IV) + Tl(II)}$$

$$\mathrm{V(II) + Tl(II) \rightarrow V(III) + Tl(I)}$$

$$\mathrm{V(III) + Tl(II) \rightarrow V(IV) + Tl(I)}$$

There is no direct reaction between thallium(I) and cerium(IV) in perchlorate media at 25°. The reaction proceeds, however, in the presence of chloride ion but the general rate behaviour indicates that an indirect route is followed and that this involves the radical ion, Cl_2^-.[39] The suggested mechanism is

$$\mathrm{CeCl^{3+} + Cl^- \underset{}{\overset{slow}{\rightleftharpoons}} Ce(III) + Cl_2^-}$$

$$\mathrm{(or\ CeCl_2^{2+} \overset{slow}{\rightleftharpoons} Ce(III) + Cl_2^-)}$$

$$\mathrm{Tl^+ + Cl_2^- \rightarrow Tl(II) + 2Cl^-}$$

$$\mathrm{Tl(II) + Cl_2^- \rightarrow Tl(III) + 2Cl^-}$$

The rate is first-order in Ce(IV) concentration but zero-order in Tl(I). A similar mechanism has been postulated in the case of the oxidation of bromide ion by Ce(IV).[40] The latter reaction was studied in acidic sulphate media and the slow steps were assigned as

$$\mathrm{Ce(SO_4)_2Br^- + Br^- \rightarrow Ce(SO_4)_2^- + Br_2^-}$$

$$\mathrm{Ce(SO_4)_2Br^- \rightarrow Ce(SO_4)_2^- + Br}$$

with a rapid pre-equilibrium

$$\mathrm{Ce(SO_4)_3^{2-} + Br^- \rightleftharpoons Ce(SO_4)_2Br^- + SO_4^{2-}}$$

In this study chloride ion enhanced the rate, an effect attributable to the formation of Cl_2^-.

The rate data for the oxidation of Tl(I) by Ce(IV) in 6·2 M nitric acid at 54° indicate the following mechanism

$$2\text{Ce(IV)} \rightleftharpoons [\text{Ce(IV)}]_2$$

$$\text{Ce(III)} + \text{Ce(IV)} \rightleftharpoons [\text{Ce(III)}.\text{Ce(IV)}]$$

$$\text{Ce(IV)} + \text{OH}^- \rightleftharpoons \text{Ce(III)} + \text{OH}$$

$$\text{Tl(I)} + \text{OH} \rightleftharpoons \text{Tl(II)} + \text{OH}^-$$

$$\text{Tl(I)} + \text{Ce(IV)} \rightleftharpoons \text{Tl(II)} + \text{Ce(III)}$$

$$\text{Tl(II)} + \text{Ce(IV)} \rightarrow \text{Tl(III)} + \text{Ce(III)}$$

where $[\text{Ce(IV)}]_2$ represents a dimeric species and [Ce(III).Ce(IV)] is a mixed dimer of Ce(III) and Ce(IV).[41] An alternative radical-producing step is

$$\text{Ce(IV)} + \text{NO}_3^- \rightleftharpoons \text{Ce(III)} + \text{NO}_3$$

Cobalt(III) oxidizes Tl(I) in a manner resembling the Tl(III)-oxidation of iron(II)[42]

$$\text{Co(III)} + \text{Tl(I)} \rightleftharpoons \text{Co(II)} + \text{Tl(II)}$$

$$\text{Co(III)} + \text{Tl(II)} \rightarrow \text{Co(II)} + \text{Tl(III)}$$

The initial presence of high concentrations of Co(II) produces a marked reduction in rate. Owing to the instability of Co(III) in perchloric acid solutions the kinetics of the reaction were studied using an excess of Tl(I) over Co(III), the disappearance of the latter being followed by titration.

The kinetics of the reaction between Ag(II) and Tl(I) in 6·2 M nitric acid at 26° have been interpreted as follows[43]

$$\text{Ag(II)} + \text{NO}_3^- \rightleftharpoons \text{Ag(I)} + \text{NO}_3$$

$$\text{NO}_3 + \text{Tl(I)} \rightleftharpoons \text{NO}_3^- + \text{Tl(II)}$$

$$\text{Ag(II)} + \text{Tl(II)} \rightarrow \text{Ag(I)} + \text{Tl(III)}$$

The kinetic data were found to be incompatible with the scheme

$$2\text{Ag(II)} \rightleftharpoons \text{Ag(I)} + \text{Ag(III)}$$

$$\text{Ag(III)} + \text{Tl(I)} \rightarrow \text{Ag(I)} + \text{Tl(III)}$$

or the scheme

$$Ag(II) + Tl(I) \rightleftharpoons Ag(I) + Tl(II)$$

$$Ag(II) + Tl(II) \rightarrow Ag(I) + Tl(III)$$

although both are in agreement with the prevailing rate law

$$\text{rate} = k[Ag(II)]^2[Tl(I)]/[Ag(I)]$$

When chromium(II) perchlorate is oxidized in aqueous solution by molecular oxygen, a binuclear species of Cr(III) is formed exclusively.[11a] The same species results from the action of thallium(III). One-equivalent oxidants, e.g., Fe(III), Cu(II), Cl_2, and Br_2, form only the hexaaquochromium(III) ion:

$$Cr^{2+} + \text{one-equivalent oxidant} \rightarrow Cr^{3+} \quad \text{(mononuclear)}$$

$$Cr^{2+} + \text{two-equivalent oxidant} \rightarrow Cr(IV)$$

$$Cr^{2+} + Cr(IV) \rightarrow \text{binuclear species}$$

Ion-exchange studies indicate the structure of the binuclear species to comprise either an oxo bridge, $[(H_2O)_5Cr—O—Cr(H_2O)_5]^{4+}$ or a diol bridge,

$$\left[(H_2O)_4Cr\begin{matrix} H \\ O \\ O \\ H \end{matrix}Cr(H_2O)_4\right]^{4+}$$

The use of oxygen-18 has been effective in deciding which of these forms represents the binuclear product.[11b] The addition of $H_2{}^{18}O$ to a solution of the latter results in the exchange of 5 oxygen atoms per chromium atom. This result is expected for a diol-bridged structure. An average of 5·5 oxygen atoms per chromium atom would exchange with labelled solvent if the structure were an oxo-bridge.

Incidentally, the formation of dimeric chromium(III) can be useful as a diagnostic test in other systems. For example, the reaction between neptunium(V) and chromium(II) was known, from kinetic evidence, to proceed via a Np(III) intermediate.[44] Two routes are possible: either a two-equivalent reduction as the initial step

$$Np(V) + Cr(II) \rightarrow Np(III) + Cr(IV)$$

$$Cr(II) + Cr(IV) \rightarrow [Cr(III)]_2$$

$$Np(III) + Np(V) \rightarrow 2Np(IV)$$

or a one-equivalent primary step

$$Np(V) + Cr(II) \rightarrow Np(IV) + Cr(III)$$

$$Np(IV) + Cr(II) \rightarrow Np(III) + Cr(III)$$

$$Np(V) + Np(III) \rightarrow 2Np(IV)$$

Both schemes are consistent with the overall stoichiometry

$$Np(V) + Cr(II) \rightarrow Np(IV) + Cr(III)$$

but the absence of dimeric chromium(III) in the product solution (together with other observations) invalidates the first scheme.

Reactions of mercury(I) and mercury(II)

Since mercury(I) is a dimeric species, $[Hg(I)]_2$, its oxidation to mercury(II) requires the overall transfer of two electrons. The oxidation of mercury(I) by cobalt(III)

$$[Hg(I)]_2 + 2Co(III) \rightarrow 2Hg(II) + 2Co(II)$$

is stoichiometric under conditions of either excess mercury(I) or equivalent concentrations of mercury(I) and cobalt(III).[45] When Co(III) is present in large excess the reaction is non-stoichiometric as a result of the slow reduction of Co(III) by water. The presence of a large excess of Co(II) or Hg(II), the products of reaction, have no influence on the rate and the simple second-order rate law is maintained. Similarly the rate is unaffected by variation of the concentration of perchlorate ion at constant acidity. This last result contrasts with the behaviour shown in the mercury(I) + thallium(III) system where an increase of ClO_4^- concentration leads to a reduction in rate (the existence of the complex $Hg_2ClO_4^+$ has been inferred from this result).

The interpretation given to the kinetics of the mercury(I) + cobalt(III) reaction is that the initial step is either

$$[Hg(I)]_2 + Co(III) \rightarrow Hg(II) + Hg(I) + Co(II)$$

or $$[Hg(I)]_2 + Co(III) \rightarrow Hg_2^{3+} + Co(II)$$

followed by

$$Hg(I) + Co(III) \rightarrow Hg(II) + Co(II)$$

or $$Hg_2^{3+} + Co(III) \rightarrow 2Hg(II) + Co(II)$$

It is noteworthy that the different kinetics displayed by the mercury(I) + thallium(III) reaction (i.e., an inverse dependence on Hg(II) concentration) indicate clearly that Tl(III) is incapable of oxidizing mercury(I) directly and that Hg(O), present in equilibrium with mercury(I) and Hg(II), is the oxidized species. The Ce(IV) + Hg(I)[46] and Co(III) + Hg(I) reactions show a close resemblance: the effect of metal-ion catalysts on the former reaction is discussed on p. 135.

The oxidation of mercury(I) by manganese(III) has been studied in aqueous perchloric acid.[47] Mn(III) is usually stabilized by using strong ($\geqslant$5 M) sulphuric acid as the medium or as the oxalato complex. Perchlorate possesses no such stabilizing ability, but reasonably stable solutions of Mn(III) perchlorate were prepared from acid permanganate and excess Mn(II), the latter suppressing the disproportionation

$$2\text{Mn(III)} \rightleftharpoons \text{Mn(II)} + \text{Mn(IV) (as MnO}_2\text{)}$$

The kinetic expression

$$-\mathrm{d}[Mn(III)]/\mathrm{d}t = k_1[Mn(III)][Hg(I)_2]/[Hg(II)] + k_2[Hg(I)_2][Mn(III)]^2/[Mn(II)]$$

applies where $[\text{Hg(I)}]_2$ represents Hg_2^{2+} (together with $Hg_2ClO_4^{+}$). The observed kinetics are in agreement with a mechanism including two rapid pre-equilibria and two slow steps:

$$[\text{Hg(I)}]_2 \underset{}{\overset{K_1}{\rightleftharpoons}} \text{Hg(O)} + \text{Hg(II)} \qquad (1)$$

$$2\text{Mn(III)} \overset{K_2}{\rightleftharpoons} \text{Mn(IV)} + \text{Mn(II)} \qquad (2)$$

$$\text{Mn(III)} + \text{Hg(O)} \xrightarrow{k'} \text{Mn(II)} + \text{Hg(I)} \quad \text{slow} \qquad (3)$$

$$\text{Mn(III)} + \text{Hg(I)} \rightarrow \text{Mn(II)} + \text{Hg(II)} \quad \text{fast} \qquad (4)$$

$$\text{Mn(IV)} + [\text{Hg(I)}]_2 \xrightarrow{k''} \text{Mn(II)} + 2\text{Hg(II)} \quad \text{slow} \qquad (5)$$

It will be observed that Mn(III) is capable of oxidizing both Hg(O) *and* Hg(I) whereas Tl(III) oxidizes only the former and Co(III) the latter. In this scheme Hg(I) is present as the Hg^+ ion and Mn(IV) as Mn^{4+} or MnO^{2+}. Step (5) may proceed via two consecutive one-equivalent stages involving Hg_2^{3+}, or Hg(I) and Hg(II). The observed constants, k_1 and k_2, in the rate equation are identified as $k'K_1$ and $k''K_2$, respectively. Other mechanisms tested and found inadequate were (*a*) the reaction of Mn(III) with $[Hg(I)]_2$, and Mn(IV) with Hg(O), and (*b*) steps (1) to (4) above with a back-reaction in step (3).

The rate of reaction between mercury(II) and vanadium(III)

$$2Hg(II) + 2V(III) \rightarrow [Hg(I)]_2 + 2V(IV)$$

is described to a large extent by the expression

$$\frac{-d[V(III)]}{dt} = \frac{[Hg(II)][V(III)]^2}{k[V(IV)] + k'[V(III)]}$$

under the conditions of 0·20 M hydrogen-ion concentration and $\mu = 3{\cdot}0$ M at 15°.[36] The following mechanism has been advanced

$$Hg(II) + V(III) \rightleftharpoons Hg(I) + V(IV)$$

$$Hg(I) + V(III) \rightarrow Hg(O) + V(IV)$$

$$Hg(II) + Hg(O) \xrightarrow{\text{fast}} [Hg(I)]_2$$

Reactions of uranium(IV)

Uranium(IV) is oxidized quantitatively to U(VI) by Fe(III) in dilute perchlorate solutions[48]

$$U(IV) + 2Fe(III) \rightarrow U(VI) + 2Fe(II)$$

At constant acidity and ionic strength the reaction is first-order in both reactants, a result compatible with either of the schemes

$$\begin{aligned} U(IV) + Fe(III) &\xrightarrow{\text{slow}} U(V) + Fe(II) \\ U(V) + Fe(III) &\xrightarrow{\text{fast}} U(VI) + Fe(II) \end{aligned} \qquad (1a)$$

$$\begin{aligned} U(IV) + Fe(III) &\xrightarrow{\text{slow}} U(V) + Fe(II) \\ 2U(V) &\xrightarrow{\text{fast}} U(VI) + U(IV) \end{aligned} \qquad (1b)$$

Of these, scheme (1a) is considered to be the most likely. Scheme (1b) is rejected on the grounds that, at the low concentrations of U(V) involved, the destruction of U(V) by Fe(III) would override the possibility of disproportionation. It should be noted that another possibility exists

$$\begin{aligned} U(IV) + Fe(III) &\xrightarrow{\text{slow}} U(VI) + Fe(I) \\ Fe(I) + Fe(III) &\xrightarrow{\text{fast}} 2Fe(II) \end{aligned} \qquad (2)$$

In this instance the reaction proceeds through the intervention of Fe(I) formed in the initial two-equivalent step. In principle the schemes might be kinetically distinguishable if the reverse of their slow steps were appreciable. If such were the case then added Fe(II) should retard the reaction in scheme (1), and in (2) the reaction would be inhibited by U(VI). Unfortunately the effect of the reaction products was apparently not investigated. An inverse dependence of the rate on hydrogen-ion concentration (to the power −1·8) suggests that the rate-determining step may be visualized as consisting of a number of competitive reactions of hydrolysed species of both reactants, predominantly $U^{4+} + Fe(OH)_2^{+}$ and $U(OH)_2^{2+} + Fe^{3+}$.

Reactions of plutonium(VI)

The kinetics of the reduction of plutonium(VI) by titanium(III) were examined by recording the absorbance of PuO_2^{2+} at 830 mμ as a function of time.[49] Pu(VI) solutions were prepared by dissolution of the metal in perchloric acid, followed by oxidation with ozone; Ti(III) solutions, prepared by dissolving the metal hydride in acid, were stored under helium. Since the reaction of Pu(V) (as PuO_2^{+}) and Ti(III) was too rapid to measure, the overall reduction of Pu(VI) was treated as comprising two consecutive second-order processes represented by

$$\mathrm{Pu(VI)} + 2\mathrm{Ti(III)} \xrightarrow{k_1} \mathrm{Pu(IV)} + 2\mathrm{Ti(IV)}$$

$$\mathrm{Pu(IV)} + \mathrm{Ti(III)} \xrightarrow{k_2} \mathrm{Pu(III)} + \mathrm{Ti(IV)}$$

The reaction between Pu(III) and Pu(VI) was insignificant. The form of rate law found for the disappearance of Pu(VI)

$$-\mathrm{d}[Pu(VI)]/\mathrm{d}t = k_1[Pu(VI)][Ti(III)]/[H^+]$$

suggests that the rate-controlling step may be

$$\mathrm{PuO_2^{2+} + TiOH^{2+} \rightarrow PuO_2^{+} + TiO^{2+} + H^+}$$

Using a value for k_2 of 65·5 s^{-1} at 25° (determined by a separate study, see below), the kinetic data were treated iteratively by a computer method to evaluate k_1. At 25° in 1 M perchloric acid, k_1 was calculated as 108 s^{-1}. Added chloride ion has no effect on the rate of reaction (an important result since slow reduction occurs of the perchlorate medium by Ti(III)). It is not clear why the reduction of PuO_2^{2+} should be more difficult to accomplish than the reduction of

PuO_2^+. The rapidity of the Pu(V) + Ti(III) reaction contrasts also with the slow reduction of Pu(V) by V(III).

The kinetics of the Pu(IV) + Ti(III) system were examined by making spectrophotometric measurements at 469 mμ, a wavelength where Pu^{4+} and $PuOH^{3+}$ are the principal absorbing species and Pu(III), Ti(III), and Ti(IV) make only a small contribution to the absorbance.[50] Pu(IV) solutions were obtained from blue Pu(III) stock solutions (prepared by dissolving the metal in acid) by oxidation with dichromate. The inverse dependence of the rate on hydrogen-ion concentration is believed to be due to the participation of a hydroxo species in the slow step: either

$$PuOH^{3+} + Ti^{3+} \rightarrow Pu^{3+} + TiO^{2+} + H^+$$

or, less probably

$$Pu^{4+} + TiOH^{2+} \rightarrow Pu^{3+} + TiO^{2+} + H^+$$

Chloride ion was demonstrated to have no effect on the rate.

Reactions of lead(IV) in acetic acid

Considering the volume of work done on redox reactions in aqueous solution it is remarkable that similar processes occurring in non-aqueous media have aroused little interest. Apart from an investigation of the reactions between lead(IV) and cerium(III), and lead(IV) and cobalt(II) in a solvent of anhydrous acetic acid, few detailed kinetic studies exist.

The oxidation of Ce(III) acetate by Pb(IV) acetate proceeds quantitatively in anhydrous acetic acid

$$Pb(IV) + 2Ce(III) \rightarrow Pb(II) + 2Ce(IV)$$

and has been followed spectrophotometrically by observing the appearance of Ce(IV) as a function of time at 400 mμ, a wavelength where the other species do not absorb.[51] A practical difficulty encountered was the photochemical instability of Ce(IV) acetate. With Pb(IV) in excess over Ce(III) the reaction is accurately first-order in each reactant over the temperature range 30–47°. Because of the possibility of an equilibrium being set up between reactants and products, the effect of added products was studied. The rate is unaffected by the presence of Ce(IV) at a concentration equal to that of the initial Ce(III) concentration. Similarly Pb(II) acetate has no effect on the rate (unless present in very large excess when a slight increase is detected). Since both Ce(IV) and Pb(II) fail to retard the reaction, the overall mechanism is probably

$$\text{Pb(IV)} + \text{Ce(III)} \xrightarrow{\text{slow}} \text{Pb(III)} + \text{Ce(IV)}$$

$$\text{Pb(III)} + \text{Ce(III)} \xrightarrow{\text{fast}} \text{Pb(II)} + \text{Ce(IV)}$$

No direct evidence was culled for the existence of Pb(III) but the observed kinetics seem to require its postulation. The reaction is concluded to proceed via ionic species; the presence of sodium acetate has a marked accelerating influence on the rate of reaction and under these conditions the ionic participants are probably $Ce(OAc)_4^-$, $Pb(OAc)_5^-$ and $Pb(OAc)_6^{2-}$. Confirmatory evidence for such species was obtained from ion migration experiments and, for Ce(III), from spectrophotometry.

The oxidation of Co(II) acetate by Pb(IV) acetate in acetic acid

$$\text{Pb(IV)} + 2\text{Co(II)} \rightarrow \text{Pb(II)} + 2\text{Co(III)}$$

is much more complicated than the Pb(IV) + Ce(III) reaction.[52] The theoretical stoichiometry is observed in the anhydrous solvent and in the presence of sodium acetate and Pb(II) acetate. But the addition of water reduces the stoichiometry as shown in Fig. 3.4. Methanol also causes a marked reduction in the stoichiometry (from 2·0 in pure acetic acid to 1·1 for 10 per cent v/v methanol–acetic acid). Since methanol and water do not bring about the decomposition of Pb(IV) acetate or of Co(III) acetate under these conditions, it is concluded that these additives must be reacting with a transient intermediate, thus causing the final concentration of Co(III) to be decreased. Whilst the order of the reaction with respect to Pb(IV) is unity under all conditions, the order with respect to Co(II) is non-integral and can be varied by the addition of Pb(II) (without Pb(II) present the order is 1·5). Added Co(III) has no effect on the rate but the addition of a large excess of Pb(II) causes a pronounced retardation.

A reaction scheme providing an explanation for most of the experimental observations is one involving dimeric species of cobalt(II), with Pb(III) and Co(IV) as reactive entities:

$$\text{Co(II)} + \text{Co(II)} \rightleftharpoons [\text{Co(II)}]_2$$

$$\text{Pb(IV)} + [\text{Co(II)}]_2 \rightarrow \text{Pb(II)} + 2\text{Co(III)}$$

$$\text{Pb(IV)} + [\text{Co(II)}]_2 \rightarrow \text{Pb(III)} + \text{Co(III)} + \text{Co(II)}$$

$$\text{Pb(IV)} + [\text{Co(II)}]_2 \rightarrow \text{Pb(II)} + \text{Co(IV)} + \text{Co(II)}$$

$$\mathrm{Pb(IV) + Co(II) \rightleftharpoons Pb(II) + Co(IV)}$$
$$\mathrm{Pb(IV) + Co(II) \rightarrow Pb(III) + Co(III)}$$
$$\mathrm{Co(IV) + Co(II) \rightarrow 2Co(III)}$$
$$\mathrm{Pb(III) + Co(II) \rightarrow Pb(II) + Co(III)}$$

The +4 oxidation state of cobalt has to be introduced to account for the retardation brought about on the addition of excess Pb(II). The

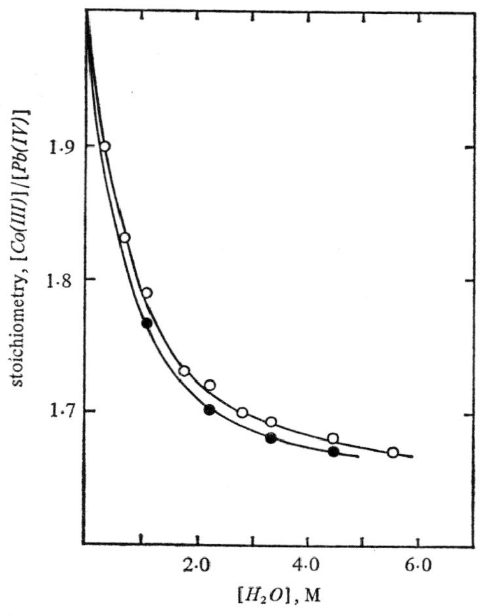

Fig. 3.4. Pb(IV) + Co(II) reaction in acetic acid: effect of water on the stoichiometry of the reaction at temperatures of 23° (denoted by ○) and 37° (denoted by ●). From D. Benson, P. J. Proll, L. H. Sutcliffe, and J. Walkley *Disc. Faraday Soc.,* 1960, **29**, 60.

postulation of a reactive dimer of cobalt(II) acetate is necessary to the rate law and is corroborated by spectrophotometric and solubility measurements. The effect of water on the stoichiometry is attributed to the direct reaction of water with Co(IV), possibly by

$$\mathrm{Co(IV) + H_2O \rightarrow Co(II) + 2H^+ + \tfrac{1}{2}O_2}$$

thereby eliminating the step

$$\mathrm{Co(IV) + Co(II) \rightarrow 2Co(III)}$$

in the reaction sequence and causing a reduction in the stoichiometry. Methanol has an effect similar to that of water, but the lower value obtained for the stoichiometric ratio $[Co(III)]/Pb(IV)]$ may be due to its reaction with Pb(III). That ionic species play an important part in the reaction is indicated by the enhancement of the rate caused by sodium acetate, a result which cannot be assigned to an inert salt effect since sodium perchlorate has a retarding influence.

Reactions of tin(II)

The existence of tin(III) in reacting systems in which Sn(II) is being oxidized can be demonstrated by its ability to reduce trioxalatocobaltate(III) to the simple Co(II) ion.[53] This complex ion is a suitable substrate because (*a*) it is insufficiently strong as an oxidizing agent to react with Sn(II) or the reduced form of the other reactant, (*b*) the product Co(II) is incapable of being reoxidized, and (*c*) low concentrations of Co(III) complexes are readily detectable spectrophotometrically. By this method Sn(III) is shown to be present as a transient intermediate in the reactions of Sn(II) with Ce(IV), Fe(III), Cr(VI), and Mn(VII). However, Sn(III) appears not to be present (or at least is not detectable) in the reactions of Sn(II) with Hg(II), Tl(III), hydrogen peroxide, iodine, and bromine.

Summary

Five distinct mechanisms are possible for non-complementary reactions between a one-equivalent oxidizing or reducing agent A and a two-equivalent reductant or oxidant B.

(*a*) One-step termolecular process:

$$2A^{m+} + B^{n+} \rightarrow 2A^{(m+1)+} + B^{(n-2)+}$$

$$\text{rate} = k[A^{m+}]^2[B^{n+}]$$

Examples: $2Fe(II) + O_2$, $2Pu(III) + O_2$, $2Ag(I) + H_2$.

(*b*) One-step bimolecular process involving a dimeric species of reductant:

$$(A^{m+})_2 + B^{n+} \rightarrow 2A^{(m+1)+} + B^{(n-2)+}$$

$$\text{rate} = k[A^{m+}]^2[B^{n+}]$$

Example: $[Co(II)]_2 + Pb(IV)$, in acetic acid.

(*c*) Bimolecular process with initial one-equivalent change:

$$A^{m+} + B^{n+} \underset{}{\overset{\text{slow}}{\rightleftharpoons}} A^{(m+1)+} + B^{(n-1)+}$$

$$A^{m+} + B^{(n-1)+} \xrightarrow{\text{fast}} A^{(m+1)+} + B^{(n-2)+}$$

$$\text{rate} = k[A^{m+}][B^{n+}]$$

Reaction retarded by $A^{(m+1)+}$ product if back reaction is significant.

Examples: Fe(II) + Tl(III) (retarded by Fe(III), no effect by Tl(I)), V(IV) + Tl(III) (retarded by V(V), no effect by Tl(I)), V(III) + Tl(III) (V(III) + V(V) → 2V(IV) too slow, therefore not mechanism (*d*))

or

$$A^{m+} + B^{n+} \overset{\text{slow}}{\rightleftharpoons} A^{(m-1)+} + B^{(n+1)+}$$

$$A^{m+} + B^{(n+1)+} \xrightarrow{\text{fast}} A^{(m-1)+} + B^{(n+2)+}$$

$$\text{rate} = k[A^{m+}][B^{n+}]$$

Reaction retarded by $A^{(m-1)+}$ product if back reaction is significant.

Examples: Co(III) + Tl(I) (retarded by Co(II), no effect by Tl(III)), Mn(III) + Hg(O) (in Mn(III) + Hg(I) reaction), Fe(III) + U(IV) (no evidence for Fe(I)), V(III) + Hg(II) (retarded by V(IV), extended form of rate law applies), Ce(IV) + Sn(II) (Sn(III) detected), Fe(III) + Sn(II) (Sn(III) detected).

(*d*) Bimolecular process with initial two-equivalent change:

$$A^{m+} + B^{n+} \overset{\text{slow}}{\rightleftharpoons} A^{(m+2)+} + B^{(n-2)+}$$

$$A^{m+} + A^{(m+2)+} \xrightarrow{\text{fast}} 2A^{(m+1)+}$$

$$\text{rate} = k[A^{m+}][B^{n+}]$$

Reaction retarded by $B^{(n-2)+}$ product if back reaction is significant.

Example: Co(II) + Pb(IV), in acetic acid (retarded by Pb(II) but not by Co(III)).

(*e*) Bimolecular process: initial disproportionation

$$2A^{m+} \overset{\text{slow}}{\rightleftharpoons} A^{(m-1)+} + A^{(m+1)+}$$

$$A^{(m-1)+} + B^{n+} \xrightarrow{\text{fast}} A^{(m+1)+} + B^{(n-2)+}$$

$$\text{rate} = k[A^{m+}]^2$$

Reaction retarded by $A^{(m+1)+}$ product. No certain examples for reactions between ions of different metals. However, the Ag(I) + Ag(II) exchange has a rate law, rate $= k[Ag(II)]^2$, and probably proceeds via the disproportionation of Ag(II), i.e., $2Ag(II) \rightleftharpoons Ag(I) + Ag(III)$ (the direct reaction $Ag(I) + Ag(II) \rightarrow Ag(II) + Ag(I)$ is very much slower).[54]

Two-equivalent–two-equivalent reactions

The oxidation of uranium(IV) by thallium(III)

The rate law obtained for the reaction

$$U(IV) + Tl(III) \rightarrow U(VI) + Tl(I)$$

is

$$-d[U(IV)]/dt = [U^{4+}][Tl^{3+}](k_1[H^+]^{-1} + k_2[H^+]^{-2})$$

implying that no direct reaction takes place between U^{4+} and Tl^{3+} but that reaction occurs via two simultaneous paths which show an inverse dependence on hydrogen-ion concentration.[55] The kinetics are interpreted as indicating the participation of hydrolysed species (UOH^{3+} and/or $TlOH^{2+}$). This is a necessary requirement since changes in coordination occur from U(IV) to U(V) (UO_2^{+}) and U(VI) (UO_2^{2+}). It is significant that the addition of U(VI) and of Tl(I) have no effect on the rate. The addition of chloride inhibits the reaction whereas sulphate ion enhances the rate. It is concluded that complexing of Tl^{3+} occurs and that the complexes so formed have reactivities radically different to that of the uncomplexed ion. Parallel behaviour is observed in the Tl(I) + Tl(III) exchange[56] and in the Tl(III) + Fe(II) reaction. The rate is unaffected by the addition of those cations, Cu(II), Ag(I), and Hg(II), which have a pronounced effect on the rate of oxidation of U(IV) by oxygen. The general features of the kinetics are more in accord with reaction via a single two-electron step rather than via successive one-electron changes: (*a*) the reaction is markedly faster than that between Fe(II) and Tl(III) which is generally considered to proceed through one-electron steps, (*b*) the reaction shows no abnormality in activation parameters which is characteristic of processes (e.g., Fe(II) + Tl(III)) occurring through the intervention of Tl(II), and (*c*) were the primary step to involve Tl(II)

$$U(IV) + Tl(III) \rightarrow U(V) + Tl(II) \tag{1}$$

then a chain mechanism would be expected to contribute to the overall reaction (as is observed in the oxidation of U(IV) by oxygen)

$$U(IV) + Tl(II) \rightarrow U(V) + Tl(I)$$

$$U(V) + Tl(III) \rightarrow U(VI) + Tl(II)$$

There is no such indication, particularly since the rate is insensitive to the presence of metallic ions like Cu(II). It is not possible to decide whether reaction (1) occurs initially, U(V) and Tl(II) reacting instantly before they have time to diffuse out of their solvent cages, or whether an instantaneous transfer of two electrons takes place. There are striking differences between the rate in aqueous perchloric acid and 75 per cent methanol–aqueous perchloric acid solutions.[57] In the mixed solvent the reaction is markedly catalysed by Cu(II) and Hg(II) and inhibited by Ag(I). These results strongly suggest a step-wise process involving U(V) and Tl(II) analogous to that advanced for the oxidation of U(IV) by oxygen (p. 138)

$$U(IV) + Tl(III) \rightarrow U(V) + Tl(II)$$

$$U(V) + Cu(II) \rightarrow U(VI) + Cu(I)$$

$$Tl(II) + Cu(I) \rightarrow Tl(I) + Cu(II)$$

Ag(I) is believed to exert an inhibitory effect, as in the $U(IV) + O_2$ reaction, by bringing about a chain-breaking step with U(V).

To summarize, there is evidence to suggest that a +2 oxidation state of thallium participates in the reactions Tl(III) + Fe(II), Tl(III) + V(IV), Tl(III) + V(III), Co(III) + Tl(I), and Ag(II) + Tl(I); but not in the reactions Tl(III) + Cr(II), Tl(III) + Hg(I), or Tl(I) + Tl(III).

The oxidation of mercury(I) by thallium(III)

Cerium(IV), although a strong oxidizing agent, reacts only slowly with $[Hg(I)]_2$ whereas thallium(III), a two-equivalent oxidant, reacts rapidly

$$[Hg(I)]_2 + Tl(III) \rightarrow 2Hg(II) + Tl(I)$$

The rate is unaffected by the presence of Tl(I) but is inhibited by Hg(II), the rate law being[58]

$$-\mathrm{d}[Hg(I)_2]/\mathrm{d}t = k'[Hg(I)_2][Tl(III)]/[Hg(II)]$$

The rate of disappearance of mercury(I) was followed spectrophotometrically at 236 mμ. The observed rate constant, k', is found to be inversely proportional to hydrogen-ion and perchlorate-ion concentrations. The mechanism proposed accounts very adequately for the observed kinetics

$$\left.\begin{array}{rcl} Hg_2^{2+} + ClO_4^- & \underset{}{\overset{K_1}{\rightleftharpoons}} & Hg_2ClO_4^+ \\ Hg_2^{2+} & \underset{}{\overset{K_2}{\rightleftharpoons}} & Hg^{2+} + Hg \\ Tl^{3+} + H_2O & \underset{}{\overset{K_3}{\rightleftharpoons}} & TlOH^{2+} + H^+ \end{array}\right\} \text{rapid equilibria}$$

$$Hg + TlOH^{2+} \xrightarrow{k} Hg^{2+} + Tl^+ + OH^- \quad \text{slow} \qquad (1)$$

The derived rate law

$$-\mathrm{d}[Hg(I)_2]/\mathrm{d}t = [kHg][TlOH^{2+}]$$

can be transformed into

$$\frac{-\mathrm{d}[Hg(I)_2]}{\mathrm{d}t} = \frac{kK_2K_3}{(1 + K_1[ClO_4^-][H^+])} \cdot \frac{[Hg(I)_2][Tl(III)]}{[Hg(II)]}$$

from which k' can be identified as

$$k' = \frac{kK_2K_3}{(1 + K_1[ClO_4^-][H^+])}$$

The use of determined values for $K_1 = 0{\cdot}91\ M^{-1}$, $K_2 = 5{\cdot}5 \times 10^{-9}$ M, and $K_3 = 0{\cdot}073$ M together with a k' value of $4{\cdot}2 \times 10^{-5}\ s^{-1}$, at $[H^+] = 3$ M, $[ClO_4^-] = 3$ M, allows the calculation of k as $1 \times 10^6\ M^{-1}\ s^{-1}$ at 25°. It is interesting that the kinetics show the direct reaction between Hg_2^{2+} and $TlOH^{2+}$

$$Hg_2^{2+} + TlOH^{2+} \rightarrow 2Hg^{2+} + OH^- + Tl^+$$

to be negligible, although the concentration of Hg_2^{2+} ions may be 10^5 times greater than that of mercury atoms. Chloride and bromide ions catalyse the reaction by complexing with Hg^{2+}, thereby increasing the concentration of mercury atoms by displacing the $Hg_2^{2+} \rightleftharpoons Hg^{2+} + Hg$ disproportionation step to the right. The rate-determining step (1) bears a formal resemblance to the slow step in the Tl(I) + Tl(III) exchange

$$Tl^+ + \overset{*}{T}lOH^{2+} \rightarrow Tl^{3+} + \overset{*}{T}l^+ + OH^- \qquad (2)$$

particularly as Tl^+ and Hg are isoelectronic. Step (2) is in fact 10^9 to 10^{10} times slower than (1) although the activation energies are of the same order. The marked difference in rate suggests that the charges of the reactant ions play an important role in determining the rate. This is reflected by large differences in the entropies of activation.

The oxidation of tin(II) by plutonium(VI)

A simple second-order rate law applies in the reaction between plutonium(VI) and tin(II) in mixed chloride–perchlorate media[59]

$$\mathrm{Pu(VI) + Sn(II) \rightarrow Pu(IV) + Sn(IV)}$$

Two mechanisms are possible for this system, assuming the common rate-determining step to be

$$\mathrm{Pu(VI) + Sn(II) \rightarrow Pu(V) + Sn(III)}$$

Either a second amount of Pu(VI) is reduced by Sn(III), followed by disproportionation of Pu(V)

$$\mathrm{Pu(VI) + Sn(III) \rightarrow Pu(V) + Sn(IV)}$$

$$\mathrm{2Pu(V) \rightarrow Pu(IV) + Pu(VI)}$$

or reduction of a second equivalent of Pu(VI) by Sn(III) is accompanied by the reduction of Pu(V) by Sn(II)

$$\mathrm{Pu(VI) + Sn(III) \rightarrow Pu(V) + Sn(IV)}$$

$$\mathrm{Pu(V) + Sn(II) \rightarrow Pu(IV) + Sn(III)}$$

$$\mathrm{Pu(V) + Sn(III) \rightarrow Pu(IV) + Sn(IV)}$$

The first possibility is eliminated on the grounds that the Pu(V) disproportionation is relatively slow. Similarly the slowness of reaction between Pu(V) and Sn(II) excludes the second scheme. Thus the reaction must proceed by a single two-equivalent process without the intervention of Sn(III) as an intermediate.

Multi-equivalent reactions

Reactions of chromium(VI) and chromium(III)

The kinetics of the chromium(VI) + neptunium(V) reaction

$$\mathrm{Cr(VI) + 3Np(V) \rightarrow Cr(III) + 3Np(VI)}$$

have been the subject of a careful investigation.[60] Data were obtained spectrophotometrically at wavelengths of 350 mμ (where Cr(VI) absorbs over ten times more strongly than Np(VI) while the absorbances due to Np(V) and Cr(III) can be neglected) and 980 mμ (where the absorbance is due entirely to Np(V)). Stock solutions of neptunium were prepared immediately before a kinetic run to avoid the possibility that hydrogen peroxide might be formed by radiolysis of the solutions by α-particles from the decay of ^{237}Np. The empirical rate law, quite unlike that of the Cr(VI) + Fe(II) or Cr(VI) + V(IV) systems, was determined by a computer programme, capable of producing from a set of experimental data an approximate solution for the first derivative of an unknown function. The rate expression obtained

$$\frac{-d[NpO_2{}^+]}{dt} = \frac{k[NpO_2{}^+][Cr(VI)]}{1 + k'[NpO_2{}^{2+}]/[NpO_2{}^+]} \tag{3.2}$$

has the following indefinite integral

$$t = A \log \frac{[NpO_2{}^+]}{[Cr(VI)]_0 - [NpO_2{}^+]_0/3 + [NpO_2{}^+]/3} + \frac{B}{[NpO_2{}^+]} + C \tag{3.3}$$

where the subscript refers to initial concentrations of the reactants, and the parameters A and B are given by

$$A = \frac{[Cr(VI)]_0(1 - k') - [Np(V)]_0/3 - k'[NpO_2{}^{2+}]_0/3}{k([Cr(VI)]_0 - [NpO_2{}^+]_0/3)^2}$$

$$B = \frac{k'([NpO_2{}^+]_0 + [NpO_2{}^{2+}]_0)}{k([Cr(VI)]_0 - [NpO_2{}^+]_0/3)} \tag{3.4}$$

Values for A, B, and C were then computed from the rate data (of $[Np(V)]$-time) using eq. (3.3), and fed into the set (3.4) to generate values for the rate constants, k and k'. Values of k and k' obtained in this manner were in agreement with those obtained directly from eq. (3.2). The rate law was tested by observing the constancy of k and k' over a range of initial concentrations of Cr(VI), Np(V), and Np(VI) at constant acidity and ionic strength.

The empirical rate law is in accord with a mechanism consisting of a set of one-equivalent steps

$$Cr(VI) + NpO_2{}^+ \underset{k_2}{\overset{k_1}{\rightleftharpoons}} Cr(V) + NpO_2{}^{2+}$$

$$Cr(V) + NpO_2{}^+ \underset{k_4}{\overset{k_3}{\rightleftharpoons}} Cr(IV) + NpO_2{}^{2+}$$

$$Cr(IV) + NpO_2{}^+ \xrightarrow{k_5} Cr(III) + NpO_2{}^{2+}$$

in which Cr(V) is recognized to be the kinetically-important intermediate ($k = 3k_1$ and $k' = k_2/k_3$).

The reaction of chromium(VI) with vanadium(IV) conforms to the stoichiometry

$$Cr(VI) + 3V(IV) \rightarrow Cr(III) + 3V(V)$$

Under conditions of 0·005 to 0·10 M perchloric acid and $\mu = 1{\cdot}0$ M at 25° the overall reaction can be rewritten essentially as

$$HCrO_4^- + 3VO^{2+} + H^+ \rightarrow Cr^{3+} + 3VO_2^+ + H_2O$$

No binuclear product of chromium(III) can be detected. Vanadium(V) retards the reaction and the full form of the rate law is given by[61]

$$\frac{-d[HCrO_4^-]}{dt} = \frac{[VO^{2+}]^2}{[VO_2^+]}(k[HCrO_4^-] + k'[H^+][HCrO_4^-]^2)$$

The first term of this expression is dominant at low concentrations of Cr(VI) ($\leqslant 4 \times 10^{-5}$ M) and at low acidities ($\leqslant 0{\cdot}03$ M). At higher concentrations of Cr(VI) and/or H^+ ($[HCrO_4^-][H^+] \geqslant 2 \times 10^{-6}$) the second term becomes assertive. Two possibilities exist for the reaction path described by the first term of the rate law:

either

$$2V(IV) \underset{k_2}{\overset{k_1}{\rightleftharpoons}} V(V) + V(III) \qquad \text{rapid equilibrium, } K$$

$$V(III) + Cr(VI) \xrightarrow{k_3} V(V) + Cr(IV) \qquad \text{slow} \qquad (1)$$

$$V(IV) + Cr(IV) \rightleftharpoons V(V) + Cr(III) \qquad \text{rapid}$$

or

$$V(IV) + Cr(VI) \rightleftharpoons V(V) + Cr(V) \qquad \text{rapid}$$

$$V(IV) + Cr(V) \rightarrow V(V) + Cr(IV) \qquad \text{slow} \qquad (2)$$

$$V(IV) + Cr(IV) \rightleftharpoons V(V) + Cr(III) \qquad \text{rapid}$$

Both mechanisms are consistent with the kinetics in that they lead to the same composition of transition state in the rate-determining step, i.e., $[HCrO_4V^{2+}]^{\ddagger}$ with an average oxidation number for Cr and V of 4·5 (0·5(2 × 4 + 6 − 5)). Such a transition state could arise from either a V(III) + Cr(VI) or V(IV) + Cr(V) slow stage.

Considering scheme (1), since K (estimated from electrode potential data) is 10^{-10} at 25° and k is 0·56 M^{-1} s^{-1} then k_3 is calculated as

$0{\cdot}56 \times 10^{10}$ M^{-1} s^{-1} ($k = k_3K$). The rate law requires that $k_2[V(V)] \gg k_3[Cr(VI)]$, i.e., if V(V) is to retard the reaction then the unstable intermediate V(III) must react with V(V) in preference to Cr(VI). The value of k_2, obtained from a study of the V(III) + V(V) reaction is $1{\cdot}4 \times 10^4$ M^{-1} s^{-1} at 25° and 0·02 M H^+.[62] Thus for the inequality $k_2[V(V)] \gg k_3[Cr(VI)]$ to hold it is required that $[V(V)]/[Cr(VI)] \gg k_3/k_2 \gg 4 \times 10^5$. However, the highest value of $[V(V)]/[Cr(VI)]$ achieved in practice is about 3×10^2 and scheme (1) is therefore invalidated. Scheme (2), consisting of three separate one-equivalent stages, is in line with the Cr(VI) + Fe(II) reaction (p. 174) and the Ce(IV) + Cr(III) reaction (below), all three reactions displaying the same form of rate law, and having a Cr(V) – Cr(IV) transformation as a slow step. The form of the second term of the rate law (showing a second-order dependence on Cr(VI)) implies that dichromate ion is a reactant in the first step and a dimeric Cr(V)–Cr(VI) species is the reaction intermediate in the slow step.

The reaction of Cr(VI) with Cr(II) in aqueous perchloric acid yields hexaaquochromium(III) and a polymeric species (presumably $Cr_2(OH)_2^{4+}$) in the stoichiometric ratio of 2:1.[11c] When the reaction is performed with radioactive Cr(VI) (labelled with ^{51}Cr) activity is found in both products, a result in agreement with the mechanism

$$\overset{*}{\mathrm{Cr}}(\mathrm{VI}) + \mathrm{Cr(II)} \longrightarrow \overset{*}{\mathrm{Cr}}(\mathrm{V}) + \mathrm{Cr}^{3+}$$

$$\overset{*}{\mathrm{Cr}}(\mathrm{V}) + \mathrm{Cr(II)} \xrightarrow{\sim 80\%} \overset{*}{\mathrm{Cr}}(\mathrm{IV}) + \mathrm{Cr}^{3+} \qquad \text{one-equivalent step}$$

$$\overset{*}{\mathrm{Cr}}(\mathrm{V}) + \mathrm{Cr(II)} \xrightarrow{\sim 20\%} \overset{*}{\mathrm{Cr}}{}^{3+} + \mathrm{Cr(IV)} \qquad \text{two-equivalent step}$$

$$\overset{*}{\mathrm{Cr}}(\mathrm{IV}) + \mathrm{Cr(II)} \longrightarrow \overset{*}{\mathrm{Cr}}(\mathrm{OH})_2\mathrm{Cr}^{4+}$$

$$\mathrm{Cr(IV)} + \mathrm{Cr(II)} \longrightarrow \mathrm{Cr(OH)_2Cr^{4+}}$$

In acidic sulphate media the rate of oxidation of Cr(III) to Cr(VI) by cerium(IV) shows an inverse dependence on the concentration of Ce(III) and a second-order dependence upon Ce(IV) concentration,[63] the rate law (at 25°) being given by

$$\frac{\mathrm{d}[Cr(VI)]}{\mathrm{d}t} = \frac{k[Ce(IV)]^2[Cr(III)]}{[Ce(III)]}$$

The transition state ($X^{\neq}$) in the rate-controlling step thus contains one cerium atom and one chromium atom, the average oxidation state of each atom being +4 ($X^{+} = 2Ce(IV) + Cr(III) - Ce(III)$). A sequence of three one-equivalent steps explains the kinetics

$$Ce(IV) + Cr(III) \underset{k_2}{\overset{k_1}{\rightleftharpoons}} Ce(III) + Cr(IV) \quad (3)$$

$$Ce(IV) + Cr(IV) \underset{k_4}{\overset{k_3}{\rightleftharpoons}} Ce(III) + Cr(V) \quad \text{rate-determining} \quad (4)$$

$$Ce(IV) + Cr(V) \underset{k_6}{\overset{k_5}{\rightleftharpoons}} Ce(III) + Cr(VI)$$

The concentration of chromium(IV) is fixed by equilibrium (3) and is given by

$$[Cr(IV)] = k_1[Cr(III)][Ce(IV)]/k_2[Ce(III)]$$

which on substitution into the rate expression

$$\text{rate} = k_3[Ce(IV)][Cr(IV)]$$

corresponding to the slow step (4), yields the rate law

$$\text{rate} = \frac{k_1 k_3}{k_2} \frac{[Ce(IV)]^2[Cr(III)]}{[Ce(III)]}$$

The rate constant k is thus identified as k_1k_3/k_2. The absence of a term in Ce(IV) in the denominator indicates that $k_2[Ce(III)] > k_3[Ce(IV)]$. From the form of the rate law the inequality $k_5[Ce(IV)] \gg k_4[Ce(III)$ applies also. As in the chromium(VI) + iron(II) reaction the slow step involves the interconversion of Cr(IV) and Cr(V), a change requiring an alteration in the coordination shell of the metal ion. Incidentally, it should be noted that the reaction was studied in sulphate media on grounds of expediency since the rate in perchloric acid solution was found to be inconveniently high for normal spectrophotometric measurements.

Metal-ion-catalysed reactions

Since the reaction between cerium(IV) and chromium(III) in perchloric acid is too rapid to be followed conveniently by simple techniques, it is surprising that the Co(III) + Cr(III) reaction proceeds only slowly, the rate being comparable to that of the thermal decomposition of Co(III) in 3 M perchloric acid. The reaction

$$3Co(III) + Cr(III) \xrightarrow{Ag(I)} 3Co(II) + Cr(VI)$$

is catalysed, however, by small concentrations of silver(I). The rate of reaction is found to increase on increasing the Ag(I) and Cr(III) concentrations, whilst the presence of Co(II) has a retarding influence.[64] The scheme which best fits these observations is one in which there is a rapid pre-equilibrium between Co(III) and Ag(I) producing Co(II) and Ag(II)

$$\mathrm{Co(III)} + \mathrm{Ag(I)} \underset{k_2}{\overset{k_1}{\rightleftharpoons}} \mathrm{Co(II)} + \mathrm{Ag(II)}$$

followed by the step-wise oxidation of Cr(III) by Ag(II)

$$\mathrm{Cr(III)} + \mathrm{Ag(II)} \xrightarrow{k_3} \mathrm{Cr(IV)} + \mathrm{Ag(I)}$$

$$\mathrm{Cr(IV)} + \mathrm{Ag(II)} \rightarrow \mathrm{Cr(V)} + \mathrm{Ag(I)}$$

$$\mathrm{Cr(V)} + \mathrm{Ag(II)} \rightarrow \mathrm{Cr(VI)} + \mathrm{Ag(I)}$$

Assuming that the concentrations of the intermediates Ag(II), Cr(IV), and Cr(V) are constant in the steady state, the derived rate law (see p. 9) becomes

$$\frac{\mathrm{d}[Cr(VI)]}{\mathrm{d}t} = \frac{k_1 k_3 [Cr(III)][Ag(I)][Co(III)]}{k_2[Co(II)] + 3k_3[Cr(III)]}$$

If the steps

$$\mathrm{Cr(IV)} + \mathrm{Co(III)} \rightarrow \mathrm{Cr(V)} + \mathrm{Co(II)}$$

$$\mathrm{Cr(V)} + \mathrm{Co(III)} \rightarrow \mathrm{Cr(VI)} + \mathrm{Co(II)}$$

are included then an identical rate law results. The substitution of Ag(III), produced by the disproportionation

$$2\mathrm{Ag(II)} \rightleftharpoons \mathrm{Ag(I)} + \mathrm{Ag(III)}$$

in place of Ag(II) as the reactive species is not in accord with the observed kinetics.

The effect of Ag(I) on the rate of oxidation of Fe(II) by Co(III) has been studied.[65] In the presence of excess Ag(I) and Fe(II) the reactions are

$$\mathrm{Co(III)} + \mathrm{Ag(I)} \underset{k_2}{\overset{k_1}{\rightleftharpoons}} \mathrm{Co(II)} + \mathrm{Ag(II)}$$

$$\mathrm{Ag(II)} + \mathrm{Fe(II)} \xrightarrow{k_4} \mathrm{Ag(I)} + \mathrm{Fe(III)}$$

$$\mathrm{Co(III)} + \mathrm{Fe(II)} \xrightarrow{k_5} \mathrm{Co(II)} + \mathrm{Fe(III)}$$

At high concentrations of Fe(II) the observed second-order rate constant for the disappearance of Co(III) is given by

$$k_{obs} = k_5 + k_1[Ag(I)]/[Fe(II)]$$

and a plot of k_{obs} versus $[Ag(I)]/[Fe(II)]$ allows the determination of k_5 (from the intercept) and k_1 (from the slope). The values obtained are $k_1 = 41 \pm 3\ M^{-1}\ s^{-1}$ and $k_5 = 330 \pm 7\ M^{-1}\ s^{-1}$, at 25° in 4 M perchloric acid. The oxidation of Co(II) by Ag(II) was investigated separately. Under conditions of excess Co(II) and Ag(I) the observed rate constant for the approach of Ag(II) concentration to equilibrium is

$$k_{obs} = k_1[Ag(I)] + k_2[Co(II)]$$

and a plot of k_{obs} versus $[Co(II)]$ at a constant Ag(I) concentration of $1{\cdot}16 \times 10^{-2}$ M is linear, allowing the determination of k_2 from the slope of the line ($k_2 = 1{\cdot}75 \times 10^3\ M^{-1}\ s^{-1}$). The value obtained for k_1 ($37 \pm 4\ M^{-1}\ s^{-1}$) is in good agreement with that above. Consequently a value can be assigned for the equilibrium constant of the Co(III) + Ag(I) system: $K_1 = k_1/k_2 = 2{\cdot}3 \times 10^{-2}$ in 4 M perchloric acid at 25°. K_1 can be calculated also from the standard electrode potentials of the Ag(II)–Ag(I) couple ($E° = 2{\cdot}00$ V) and the Co(III)–Co(II) couple ($E° = 1{\cdot}92 \pm 0{\cdot}02$ V) since the standard e.m.f. is given by $0{\cdot}059 \log K_1$. By this method a value of $(4 \pm 2) \times 10^{-2}$ is obtained. The rates of oxidation of cobalt(II), chromium(III), cerium(III),

Table 3.5

Rate constants for silver ion and cobalt ion reactions, in 4 M perchloric acid at 25° *

Reductant	$k_{Ag(II)}$, $M^{-1}\ s^{-1}$	$k_{Co(III)}$, $M^{-1}\ s^{-1}$	$k_{Ag(II)}/k_{Co(III)}$	$k_{Ag(I)}/k_{Co(II)}$
Co(II)	$1{\cdot}75 \times 10^3$	20	$0{\cdot}9 \times 10^2$	2
Cr(III)	$1{\cdot}5 \times 10^3$†	$\leqslant 4$	$\geqslant 4 \times 10^2$	$\geqslant 9$
Ce(III)	5×10^3	10§	5×10^2	10
V(IV)	5×10^3	$1{\cdot}37$	$3{\cdot}6 \times 10^3$	8
Fe(II)	4×10^5	$3{\cdot}3 \times 10^2$	$1{\cdot}2 \times 10^3$	3
Mn(II)	3×10^4	$1{\cdot}0 \times 10^2$†	3×10^2	7

* From ref. (65), and references cited therein.
† 3 M $HClO_4$.
§ 1 M $HClO_4$.

vanadium(IV), iron(II), and manganese(II) by Ag(II) and Co(III) are collected in Table 3.5. It is seen from Table 3.5 that the oxidations by Ag(II) are 10^2 to 10^4 times more rapid than those of Co(III). Furthermore, it is possible to calculate the relative rates of oxidation of Ag(I) and Co(II) by the oxidized form of the couple since the equilibrium constant of the Co(II) + Ag(II) system is available. For example, in the equilibria

$$\text{Fe(II)} + \text{Co(III)} \rightleftharpoons \text{Fe(III)} + \text{Co(II)}$$

and

$$\text{Fe(II)} + \text{Ag(II)} \rightleftharpoons \text{Fe(III)} + \text{Ag(I)}$$

the forward rate constants are known and so also is K_1, thus permitting the ratio of rate constants, $k_{\text{Ag(I)}}/k_{\text{Co(II)}}$, to be obtained for the reaction of Ag(I) and Co(II) with Fe(III). Inspection of the last column in Table 3.5 reveals that the reductions by Ag(I) are more rapid than reductions by Co(II). This illustrates clearly that silver ion, in general, should prove more efficient than cobalt ion as a catalyst in redox reactions, both the oxidized and the reduced forms being more highly reactive.

Silver(I) is effective as a catalyst in the oxidations of mercury(I) and thallium(I) by cerium(IV).[36] No direct reaction occurs between Ce(IV) and Ag(I) or, under the conditions of the studied kinetics, between Ce(IV) and mercury(I) or Tl(I). In the case of the Ag(I)-catalysed reaction between Ce(IV) and mercury(I)

$$2\text{Ce(IV)} + [\text{Hg(I)}]_2 \xrightarrow{\text{Ag(I)}} 2\text{Ce(III)} + 2\text{Hg(II)}$$

the rate law assumes the form

$$\begin{aligned} -\mathrm{d}[Ce(IV)]/\mathrm{d}t &= -2\mathrm{d}[Hg(I)_2]/\mathrm{d}t \\ &= 2k'[Ce(IV)] = 2k[Ce(IV)][Ag(I)] \end{aligned}$$

in the presence of a large excess of $[\text{Hg(I)}]_2$. Thus the rate of reaction is found to be independent of the concentration of mercury(I). Also, if mercury(I) is in small excess only and Ce(III) is present initially, then plots of log $[Ce(IV)]$ versus time show curvature, k' decreasing as the Ce(III) product increases. This observation is indicative of a back-reaction involving Ce(III). The mechanistic scheme

$$\text{Ce(IV)} + \text{Ag(I)} \underset{k_2}{\overset{k_1}{\rightleftharpoons}} \text{Ce(III)} + \text{Ag(II)}$$

$$\text{Ag(II)} + [\text{Hg(I)}]_2 \xrightarrow{k_3} \text{Ag(I)} + \text{Hg(I)} + \text{Hg(II)}$$

$$\text{Ce(IV)} + \text{Hg(I)} \xrightarrow{k_4} \text{Ce(III)} + \text{Hg(II)}$$

is in agreement with the kinetics since, by assuming stationary-state concentrations for Ag(II) and Hg(I), it produces the expression

$$\frac{-\mathrm{d}[Ce(IV)]}{\mathrm{d}t} = \frac{-2d[Hg(I)_2]}{\mathrm{d}t}$$

$$= \frac{2k_1k_3[Ce(IV)][Ag(I)][Hg(I)_2]}{k_2[Ce(III)] + k_3[Hg(I)_2]}$$

Under conditions of excess $[Hg(I)]_2$ the inequality $k_3[Hg(I)_2] \gg k_2[Ce(III)]$ holds and the expression reduces to one of simple second-order ($k = k_1$). The Ag(I)-catalysed oxidation of Tl(I) by Ce(IV) proceeds by a similar route

$$Ce(IV) + Ag(I) \rightleftharpoons Ce(III) + Ag(II)$$

$$Ag(II) + Tl(I) \rightarrow Ag(I) + Tl(II)$$

$$Ce(IV) + Tl(II) \rightarrow Ce(III) + Tl(III)$$

The uncatalysed reaction between iron(III) and vanadium(III) displays some unusual features not expected of a one-equivalent complementary reaction.[66] Besides the obvious step

$$Fe(III) + V(III) \xrightarrow{k_1} Fe(II) + V(IV) \qquad (1)$$

the complex rate law discloses a significant contribution from the steps

$$Fe(III) + V(IV) \underset{k_{-2}}{\overset{k_2}{\rightleftharpoons}} Fe(II) + V(V) \qquad (2)$$

$$V(V) + V(III) \xrightarrow{k_3} 2V(IV) \qquad (3)$$

In the presence of copper(II) the rate of reaction becomes independent of the concentrations of Fe(III), Fe(II), and V(IV)

$$-\mathrm{d}[Fe(III)]/\mathrm{d}t = -\mathrm{d}[V(III)]/\mathrm{d}t$$

$$= k'[V(III)]$$

$$= k[V(III)][Cu(II)]$$

A two-stage mechanism is considered appropriate

$$Cu(II) + V(III) \xrightarrow{k} Cu(I) + V(IV)$$

$$Fe(III) + Cu(I) \xrightarrow{\text{fast}} Fe(II) + Cu(II)$$

The oxidation of vanadium(III) by uranium(VI) takes place through the intervention of U(V) which disproportionates ultimately to U(IV) and U(VI). In acid solutions the equilibrium

$$\mathrm{U(VI) + V(III)} \underset{k_{-4}}{\overset{k_4}{\rightleftharpoons}} \mathrm{U(V) + V(IV)} \tag{4}$$

lies far to the left ($K \sim 6 \times 10^{-6}$ at 25°) and the reaction is very slow. The kinetics were studied by utilizing the fact that U(V) reacts rapidly with Fe(III)[67]

$$\mathrm{U(V) + Fe(III)} \xrightarrow{k_5} \mathrm{U(VI) + Fe(II)} \tag{5}$$

Thus the presence of U(VI) catalyses the reaction between V(III) and Fe(III), which in the absence of catalyst is quite slow. The kinetics of the catalysed reaction are adequately accounted for by steps (4) and (5) together with a 'background' reaction made up of (1) to (3). The rate law is complex and is given by

$$\frac{\mathrm{d}[V(IV)]}{\mathrm{d}t} = \frac{k_4[V(III)][U(VI)]}{1 + (k_{-4}/k_5)[V(IV)]/[Fe(III)]} + k_1[V(III)][Fe(III)] + \frac{k_2[V(IV)][Fe(III)]}{1 + (k_{-2}/k_3)[Fe(II)]/[V(III)]}$$

The first term of this expression relates to the catalysed reaction while the second and third terms describe the small, but not negligible, uncatalysed part of the reaction. Over the range of acidities, 0·05 to 1·67 M perchloric acid at $\mu = 2{\cdot}0$ M, there is a strong pH dependence. Plots of log k_4 versus log $[H^+]$, at constant temperature, are nearly linear with slopes close to $-1{\cdot}8$. The implication is that two activated complexes are formed, the most important of which results from the loss of two hydrogen ions

$$UO_2^{2+} + V^{3+} + H_2O \rightarrow [VO.UO_2^{3+}]^{\ddagger} + 2H^+$$

There is a small contribution also from the path

$$UO_2^{2+} + V^{3+} + H_2O \rightarrow [VOH.UO_2^{4+}]^{\ddagger} + H^+$$

A detailed treatment of the data shows that the mechanism through which the principal activated complex is formed is more in accord with consecutive than with parallel reactions. The intermediate is almost certainly of the binuclear inner-sphere type (bridged by one or more OH groups, or by an oxygen atom). It is of interest to make a comparison with the V(III) + Np(VI) and V(III) + Pu(VI) systems since all three have a common net activation process corresponding to

$$V^{3+} + MO_2^{2+} + H_2O \rightarrow [VOH.MO_2^{4+}]^{\ddagger} + H^+$$

The entropies of activation ($\Delta S^{\ddagger}$) are 3·8, –9, and –5 cal mole^{-1} deg^{-1} for the reaction of V(III) with U(VI), Np(VI),[68] and Pu(VI),[69] respectively. It seems possible, therefore, that the Np(VI) and Pu(VI) reactions involve outer-sphere activated complexes in contrast to the U(VI) + V(III) reaction which has a positive $\Delta S^{\ddagger}$.

Reactions of molecular oxygen with metal ions

A kinetic study of the oxidation of uranium(IV) by oxygen is of intrinsic interest since the overall oxidation of U(IV) requires the loss of two electrons while the reduction of oxygen involves a net gain of four electrons

$$2U^{4+} + O_2 + 2H_2O \rightarrow 2UO_2^{2+} + 4H^+$$

In practice, oxygen gas was bubbled into the reactant solution through a sintered glass plate and the reaction was followed by withdrawing samples and titrating the unreacted U(IV).[70] Since the reaction was not affected by variation in the flow rate, it was concluded that solutions were saturated with gas, and the concentration of oxygen in solution was estimated from solubility data.

The presence of hydrogen peroxide could not be detected during the course of the reaction. Addition of U(VI), the product of reaction, has no effect on the rate. Over a wide range of conditions the kinetic data fit the expression

$$-\mathrm{d}[U(IV)]/\mathrm{d}t = k[U(IV)][O_2]/[H^+]$$

The rate of reaction is greatly enhanced by the addition of copper(II) and inhibited by the addition of silver(I), the latter ion producing a well-defined induction period. These (and other) effects indicate a chain mechanism (rather than a simple two-equivalent transfer) in which U(V) (as UO_2^+) and HO_2 act as chain-carriers:

Rapid pre-equilibrium:

$$U^{4+} + H_2O \overset{K}{\rightleftharpoons} UOH^{3+} + H^+$$

Initiation:

$$UOH^{3+} + O_2 + H_2O \xrightarrow{k_1} UO_2^+ + HO_2 + 2H^+$$

Propagation:

$$UO_2^+ + O_2 + H_2O \xrightarrow{k_2} UO_2^{2+} + HO_2 + OH^-$$

$$HO_2 + UOH^{3+} + H_2O \xrightarrow{k_3} UO_2^+ + H_2O_2 + 2H^+$$

Termination:

$$UO_2^+ + HO_2 + H_2O \xrightarrow{k_4} UO_2^{2+} + H_2O_2 + OH^-$$

(possibly in competition with the disproportionation of UO_2^+)

$$2UO_2^+ + 4H^+ \rightarrow UO_2^{2+} + U^{4+} + H_2O$$

Rapid reaction:

$$U^{4+} + H_2O_2 \rightarrow UO_2^{2+} + 2H^+$$

Assuming steady-state kinetics to apply, the following rate expression, compatible with that observed, can be derived on the basis of this scheme

$$-d[U(IV)]/dt = 2k_1K(1 + (k_2k_3/k_1k_4)^{1/2})[U^{4+}][O_2]/[H^+]$$

The catalytic influence of Cu(II) is explained in terms of reactions which generate chain-carriers

$$Cu^{2+} + UOH^{3+} + H_2O \rightarrow Cu^+ + UO_2^+ + 3H^+$$

$$Cu^+ + O_2 + H^+ \rightarrow Cu^{2+} + HO_2$$

On the other hand, Ag^+ is believed to inhibit the reaction by removing UO_2^+ and HO_2

$$UO_2^+ + Ag^+ \rightarrow UO_2^{2+} + Ag$$

$$HO_2 + Ag^+ \rightarrow O_2 + H^+ + Ag$$

After all the silver(I) ion has been removed (as metallic silver) the reaction proceeds at its normal rate.

Iron(II) is oxidized by molecular oxygen at a measurable rate between 25 and 40° in perchloric acid solution

$$4Fe^{2+} + O_2 + 4H^+ \rightarrow 4Fe^{3+} + 2H_2O$$

The rate law describing the kinetics[71]

$$-d[Fe^{2+}]/dt = k[Fe^{2+}]^2[O_2]$$

preserves its first-order dependence on oxygen even under a gas pressure of 100–130 atm. The third-order rate constant, k, increases slightly with increase in pH. Peroxide complexes of Fe(II) and O_2 are thought to take part in the reaction, for example

$$Fe^{2+} + O_2 \rightleftharpoons Fe^{2+}O_2$$

$$Fe^{2+}O_2 + H_2O.Fe(H_2O)_5{}^{2+} \xrightarrow{\text{slow}} FeO_2H^{2+} + HO.Fe(H_2O)_5{}^{2+}$$

Although copper(II) ions enhance the rate, their catalytic effect is only small.

Catalysed reactions of molecular hydrogen and carbon monoxide with metal ions

Reactions of molecular hydrogen can be catalysed not only by solids in the gas phase but also by metal ions in homogeneous solution. The field of homogeneous catalytic hydrogenations has attracted a great deal of attention in recent years. One of the more spectacular developments has been the discovery of a number of homogeneous catalysts for the hydrogenation of olefins. A number of simple catalysed reactions between hydrogen and metal ions are discussed below.

Copper(II) catalyses the oxidation of hydrogen by chromium(VI),[72] the rate law being

$$\frac{-d[H_2]}{dt} = \frac{k_1 k_3[H_2][Cu^{2+}]^2}{k_2[H^+] + k_3[Cu^{2+}]}$$

where the rate constants are defined by the scheme

$$Cu^{2+} + H_2 \underset{k_2}{\overset{k_1}{\rightleftharpoons}} CuH^+ + H^+$$

$$CuH^+ + Cu^{2+} \xrightarrow{k_3} 2Cu^+ + H^+$$

These reactions are followed by regeneration of Cu^{2+} through a sequence of steps summarized by

$$3Cu^+ + Cr(VI) \xrightarrow{\text{fast}} 3Cu^{2+} + Cr(III)$$

At high ratios of Cr(VI) to Cu^{2+}, CuH^+ is oxidized directly by Cr(VI).[73] The reversible heterolytic splitting of H_2 by Cu^{2+} is considered to account also for the oxidation of Cu^+ by hydrogen atoms:[74]

$$Cu^+ + H \rightarrow CuH^+$$

$$CuH^+ + H^+ \xrightarrow{k_2} Cu^{2+} + H_2$$

Silver(I) is an efficient catalyst for the oxidation of hydrogen.[75] With Cr(VI) as substrate the rate law is more complex than the Cu(II)-catalysed reaction and the probable mechanism has as the initial steps

$$Ag^+ + H_2 \rightleftharpoons AgH + H^+$$

together with

$$2Ag^+ + H_2 \rightarrow 2AgH^+$$

AgH^+ has been cited as an intermediate in the reduction of Ag^+ by hydrogen atoms, and also in the Ag^+-catalysed oxidation of hydrogen by permanganate where the rate-determining step is thought to be[76]

$$H_2 + Ag^+ + MnO_4^- \rightarrow AgH^+ + MnO_4H^-$$

Both mercury(I) and mercury(II) oxidize hydrogen homogeneously, the reactions displaying simple second-order kinetics.[77] Hydrido-metal intermediates are ruled out on both kinetic and energetic grounds. Furthermore, the observation that the solubility of mercury in water remains unaffected by an increase in hydrogen-ion concentration suggests that HgH^+ is unstable with respect to Hg_{aq} and H^+. The activation of hydrogen by Hg_2^{2+} and Hg^{2+} probably takes place by the simple two-electron processes

$$Hg_2^{2+} + H_2 \rightarrow 2Hg_{aq} \text{ (or } Hg_2) + 2H^+$$

and

$$Hg^{2+} + H_2 \rightarrow Hg_{aq} + 2H^+$$

The analogous reductions for Cu^{2+} and Ag^+ are not favoured energetically.

The oxidation of hydrogen by iron(III) is catalysed by ruthenium(III), rhodium(III), and palladium(II). Heterolytic splitting of hydrogen has been invoked in all three cases, for example:

$$PdCl_4^{2-} + H_2 \rightarrow HPdCl_3^{2-} + HCl$$

Pentacyanocobaltate(II), $[Co(CN)_5]^{3-}$, is a catalyst for the homogeneous hydrogenation of conjugated olefins such as butadiene and styrene.[78] The catalytic function of this five-coordinated cobalt(II) complex is attributed to the formation of a hydrido derivative, $[HCo(CN)_5]^{3-}$, on uptake of hydrogen

$$2[Co(CN)_5]^{3-} + H_2 \rightleftharpoons 2[HCo(CN)_5]^{3-}$$

Reaction with butadiene follows as

$$[HCo(CN)_5]^{3-} + C_4H_6 \rightarrow [C_4H_7Co(CN)_5]^{3-}$$

$$[C_4H_7Co(CN)_5]^{3-} + [HCo(CN)_5]^{3-} \rightarrow C_4H_8 + 2[Co(CN)_5]^{3-}$$

In this respect the $[Co(CN)_5]^{3-}$ ion is unstable in aqueous solution and hydrogen is liberated spontaneously

$$2[Co(CN)_5]^{3-} + H_2O \rightarrow [HCo(CN)_5]^{3-} + [Co(CN)_5OH]^{3-}$$

$$2[HCo(CN)_5]^{3-} \rightleftharpoons 2[Co(CN)_5]^{3-} + H_2$$

Nuclear magnetic resonance measurements have confirmed the presence of a hydrido complex.

Carbon monoxide, normally a relatively unreactive reducing agent, is oxidized in aqueous solution by Hg^{2+} and MnO_4^- but, unlike molecular hydrogen, is inert to attack by Cu(II), Ag^+ and Hg_2^{2+}. The reduction of Hg^{2+}

$$2Hg^{2+} + CO + H_2O \rightarrow Hg_2^{2+} + CO_2 + 2H^+$$

which has a simple second-order rate law, is believed to take place by the insertion of CO between Hg^{2+} and a coordinated water molecule[79]

$$\text{—}Hg^{2+}OH_2 + CO \xrightarrow{\text{slow}} \left[\text{—}Hg\text{—}\overset{\overset{\displaystyle O}{\|}}{C}\text{—}OH\right]^+ + H^+$$

together with the reactions

$$\left[\text{—}Hg\text{—}\overset{\overset{\displaystyle O}{\|}}{C}\text{—}OH\right]^+ \xrightarrow{\text{fast}} Hg + CO_2 + H^+$$

$$Hg + Hg^{2+} \xrightarrow{\text{fast}} Hg_2^{2+}$$

This postulation is supported by the observation that mercury(II) acetate reacts with CO to form AcO—Hg—CO—OCH_3, a compound analogous to the proposed intermediate. The suggestion has been made that the reduction of MnO_4^- by CO (to MnO_2 in acid or neutral solutions, and to MnO_4^{2-} in basic solutions) takes place via the intermediate [—Hg—CO—$OMnO_3$].

Reactions of the hydrated electron

Although the hydrated electron was postulated as a transient species in irradiated water as early as 1953 by Platzman, real spectroscopic evidence for its existence came in 1962 when Hart and Boag[80] observed and recorded the absorption spectrum of the entity. The technique used for producing the hydrated electron was pulse radiolysis, a radiation chemistry approach similar to the photochemical technique of flash photolysis. The source of radiation was a linear accelerator generating a beam of 1·8 MeV electrons at pulses of 2 μs. This electron beam was directed onto a quartz absorption cell containing the de-aerated solution (sodium carbonate solution and pure water were used in the original experiments). The sample cell was illuminated at right angles to the electron beam by a uranium-electrode spark source, giving 4-μs sparks, which was arranged to be triggered either simultaneously with the electron pulse or after a known time delay. The spectrum of the irradiated solutions was then recorded photographically using a spectrograph. By recording spectra at different time intervals after the pulse the rate of decay of the hydrated electron was determined: e^{-}_{aq} was considerably less reactive (lifetime ~25 μs) in sodium carbonate solution than in pure water. The presence of oxygen in the irradiated media drastically reduced the intensity of absorption. Electron spin resonance as a means of characterization is, at present, impracticable in view of the transient nature of e^{-}_{aq} and the technical problems resulting from the need to irradiate samples in the spectrometer cavity. However, the species Zn^{+} and Cd^{+}, produced by the interaction of e^{-}_{aq} with Zn^{2+} and Cd^{2+} in irradiated ice, have been detected by this method.

The hydrated electron produced by pulse radiolysis has a distinctive absorption spectrum (Fig. 3.5). The wavelength of maximum absorption is 720 mμ corresponding to a molar absorptivity value of 15,800 M^{-1} cm^{-1}. The shape of the spectrum closely resembles that of the well-documented solvated electron in liquid ammonia and methylamine. That the species has unit negative charge is shown by an application of the Debye–Hückel–Brønsted equation

$$\log \frac{k}{k_0} = 1{\cdot}02\, Z_A Z_B \frac{\mu^{1/2}}{1 + \alpha \mu^{1/2}}$$

Here k is the rate constant at an ionic strength μ for the reaction between species A and B of charges Z_A and Z_B, respectively; k_0 is the rate constant at $\mu = 0$ and α is a constant whose value is dependent upon the distance of closest approach of A and B (see eqs. (1.4) to (1.8), p. 3). The rate constants for the system

$$e^{-}_{aq} + [Fe(CN)_6]^{3-} \rightarrow [Fe(CN)_6]^{4-}$$

have been studied as a function of the ionic strength, and the slope ($= Z_A Z_B$) of a plot of $\log k/k_0$ versus $\mu^{1/2}/(1 + \alpha\mu^{1/2})$ at low ionic strengths has been measured as +3, a convincing demonstration that the hydrated electron has unit charge.[81]

Hydrated electrons are slowly converted to hydrogen atoms in pure water

$$e^-_{aq} + H_2O \underset{k_2}{\overset{k_1}{\rightleftharpoons}} H + OH^-_{aq} \qquad (1)$$

The rate constants k_1 and k_2 have been calculated as 25 M^{-1} s^{-1} and 2×10^7 M^{-1} s^{-1}, respectively, at 25°. Thus the equilibrium constant

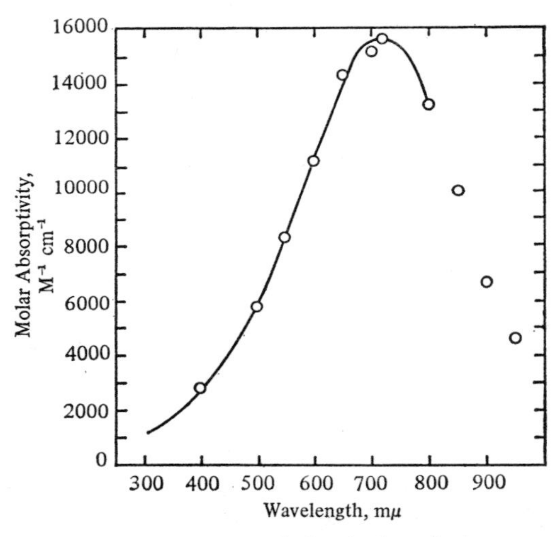

Fig. 3.5. The absorption spectrum of the hydrated electron. From M. S. Matheson, in *Solvated Electron,* Advances in Chemistry Series (ed. R. F. Gould), No. 50, p. 46, American Chemical Society, 1965.

for the reaction is $1{\cdot}25 \times 10^{-6}$ and the corresponding standard free energy change, $\Delta G°$, is 8·15 kcal $mole^{-1}$. Combination of eq. (1) with

$$H \rightarrow \tfrac{1}{2}H_2 \qquad \Delta G° = -48{\cdot}5 \text{ kcal mole}^{-1}$$

and

$$H^+ + OH^- \rightarrow H_2O \qquad \Delta G° = -21{\cdot}8 \text{ kcal mole}^{-1}$$

yields a value of −62·15 kcal $mole^{-1}$ for $\Delta G°$ of the reaction

$$e^-_{aq} + H^+ \rightarrow \tfrac{1}{2}H_2$$

This result has been used to calculate the standard oxidation potential ($E°$) of e^-_{aq}. Since $E°$ for the reaction

$$\tfrac{1}{2}H_2 \rightarrow H^+ + e^-_M$$

is zero (where e^-_M represents an electron in the normal electrochemical sense), a value of 2·7 V is obtained for $E°$ of

$$e^-_{aq} \rightarrow e^-_M$$

by making use of the relationship $\Delta G° = -nE°F$. By comparison, $E°$ for hydrogen atoms,

$$H \rightarrow H^+ + e^-_M$$

is only 2·1 V. Thus the hydrated electron is a more effective reducing agent than the hydrogen atom. Even in pure water the hydrated electron decays principally by reaction (1) with a half-life of about 230 μs.

Table 3.6

Some properties of the hydrated electron *

Wavelength of maximum absorption (λ_{max})	720 mμ (1·72 eV)
Molar absorptivity at λ_{max}	15,800 M^{-1} cm^{-1}
Molar absorptivity at 578 mμ	10,600 M^{-1} cm^{-1}
$E°(e^-_{aq} + H_3O^+_{aq} \rightarrow \frac{1}{2}H_2 + H_2O)$	2·7 V
Half-life in pure water (pH 7)	230 μs
Charge	−1
Mean radius of charge (calculated)	2·5–3·0 Å
Hydration energy (calculated)	~40 kcal (1·72 eV)

* From M. S. Matheson, in *Solvated Electron*, Advances in Chemistry Series (ed. R. F. Gould), No. 50, p. 47, American Chemical Society, 1965.

The production of a hydrated electron by radiolysis occurs by a number of successive stages. The secondary electrons produced in solution as the first products of radiolysis reach thermal equilibrium with their surroundings within 10^{-13} s. Solvation occurs within 10^{-11} s with a hydration energy of about 40 kcal $mole^{-1}$ (corresponding to the energy of the absorption maximum, 1·72 eV) and the charge is now spread out over several water molecules. It requires about 3×10^{-9} s before the hydrated electron comes to equilibrium with its ionic environment; a reaction proceeding faster than this is one involving a hydrated electron without its ionic atmosphere. After 3×10^{-9} s has elapsed the hydrated electron attains its eventual form and is accompanied by a sphere of positively-charged ions. Some properties of the hydrated electron are summarized in Table 3.6

With the exception of Mg^{2+} which reacts slowly with e^-_{aq} ($k < 10^5$

$M^{-1} s^{-1}$) the alkali and alkaline-earth metal ions do not react with the hydrated electron. This is to be expected if the high redox potential of e^{-}_{aq} is considered. All other hydrated metal ions are reduced by e^{-}_{aq} and the results of Table 3.7 for bivalent metal ions show a range of rate constants from $4{\cdot}2 \times 10^{10}$ $M^{-1} s^{-1}$ for Cr^{2+} to $7{\cdot}7 \times 10^{7}$ $M^{-1} s^{-1}$

Table 3.7

Rate constants for reactions of the hydrated electron with bivalent metal ions *

Ion	Electronic configuration	k, $M^{-1} s^{-1}$
Cr^{2+}	$3d^4$	$4{\cdot}2 \times 10^{10}$
Mn^{2+}	$3d^5$	$7{\cdot}7 \times 10^{7}$
Fe^{2+}	$3d^6$	$\sim 3{\cdot}5 \times 10^{8}$
Co^{2+}	$3d^7$	$1{\cdot}2 \times 10^{10}$
Ni^{2+}	$3d^8$	$2{\cdot}2 \times 10^{10}$
Cu^{2+}	$3d^9$	$2{\cdot}9 \times 10^{10}$
Zn^{2+}	$3d^{10}$	$1{\cdot}5 \times 10^{9}$

* From J. H. Baxendale *et al.*, *Nature Lond.*, 1964, **201**, 468, with the exception of Cr^{2+} (from ref. (85)).

for Mn^{2+}. The principal factors deciding the reactivity of a metal ion towards e^{-}_{aq} appear to be the electronic configuration of the metal ion (particularly with regard to the availability of a vacant d orbital) and the energy change on the addition of an electron. Thus in the sequence Cr^{2+}, Mn^{2+}, Fe^{2+}, Co^{2+}, Ni^{2+}, Cu^{2+}, and Zn^{2+}, the high reactivity of the $3d^4$ system of Cr^{2+} may be connected with the attainment of a stabilized half-filled d-shell on addition of an electron. Conversely, the low rate of reaction between $3d^5$ Mn^{2+} and e^{-}_{aq} may be related to the resulting destabilization of the half-filled structure, the additional electron having a spin anti-parallel to the original five spins on the metal. For $3d^9$ Cu^{2+} the gain of an electron leads to a completely-filled d shell, a particularly stable configuration: it is pertinent that Cr^{2+} and Cu^{2+} have comparable reactivities. In the case of $3d^{10}$ Zn^{2+}, since all the d orbitals are filled, the extra electron would have to be accommodated in the high energy $4s$ shell and as a result this ion reacts slowly with e^{-}_{aq}. Certainly, this approach is successful in explaining the gross features of the reactivity order but the rates for Co^{2+} and Ni^{2+} seem to be anomalously high. A comparison of the reactivities of metal ions of the same group but different period (e.g., Zn^{2+} and Cd^{2+}, Sn^{2+} and Pb^{2+}, Ni^{2+} and Pd^{2+}) reveals the larger metal ion to be the most reactive. The greater availability of vacant orbitals for ions further down the group may be the cause of their increased rate of reaction with e^-_{aq}. The most reactive trivalent ions of the lanthanide series are Sm^{3+}, Eu^{3+}, and Yb^{3+}. Significantly these are the only elements to exhibit a stable +2 oxidation state. Furthermore, their reactivities towards e^-_{aq} parallel the redox potentials of the M^{3+}–M^{2+} couples.[82]

Pulse radiolysis of aqueous solutions of Eu(III), Fe(III), Co(III), and Cr(III) leads to the production of stable products of lower oxidation number. However, reaction of e^-_{aq} (produced by a 2 μs pulse of 4 MeV electrons) with bivalent metal ions (Zn^{2+}, Mn^{2+}, Cd^{2+}, Co^{2+}, and Ni^{2+}) generates products which have a transitory existence only.[83] The absorption spectra of Zn^+, Mn^+, Cd^+, Co^+, and Ni^+ have been obtained (Fig. 3.6) and it would seem that e^-_{aq} is removed by

$$M^{2+} + e^-_{aq} \rightarrow M^+$$

before hydrogen atoms can be formed by

$$H_2O + e^-_{aq} \rightarrow H + OH^-$$

The rate of decay of these transient species follows second-order kinetics, and it is likely that their mode of disappearance is by disproportionation, for example,

$$Ni^+ + Ni^+ \rightarrow Ni^{2+} + Ni(O)$$

The absorption of the ion Zn^+ has a peak at about 300 mμ and a corresponding molar absorptivity of about 1300 M^{-1} cm^{-1}. The rates of oxidation of the transients with M^{2+} ions, for example,

$$Cd^+ + Pb^{2+} \rightarrow Cd^{2+} + Pb^+$$

have been determined by observing the rate of decay of the M^+ ion in the presence of oxidant.[84] The transients (e.g., Cd^+) were produced in the concentration range 1–5 μM by a 0·2 μs pulse of 4 MeV electrons on neutral M^{2+} solutions (2–20 mM). Oxidant (e.g., Pb^{2+}) was added in small concentrations (50–200 μM). Table 3.8 contains the results

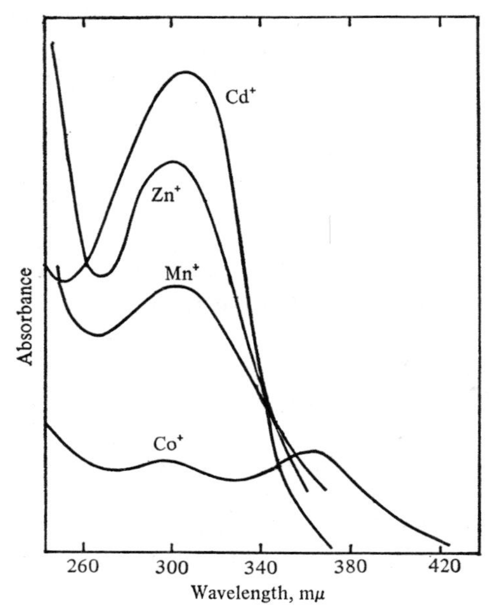

Fig. 3.6. Unstable oxidation states produced by an electron pulse on aqueous solutions of Cd^{2+}, Zn^{2+}, Mn^{2+}, and Co^{2+}. From G. E. Adams, J. H. Baxendale, and J. W. Boag, *Proc. Chem. Soc.,* 1963, 241.

Table 3.8

Rate constants for reaction of M^+ ions with bivalent metal ions*

	$10^{-8}k$, $M^{-1}\ s^{-1}$, at 18°			
Oxidant	Zn^+	Cd^+	Pb^+	Ni^+
Zn^{2+}	57†	0	0	0
Cd^{2+}	8·3	180†	0	0
Pb^{2+}	4·0	7·5	7·7†	0
Ni^{2+}	0·5	0	0	63†

* From ref. (84).

† For the decay (by disproportionation) of M^+ in the absence of added oxidant.

obtained. Since Zn^+ is oxidized by Cd^{2+}, and Cd^+ is oxidized by Pb^{2+}, the standard electrode potentials must be in the order $E°(Pb^{2+}, Pb^+) > E°(Cd^{2+}, Cd^+) > E°(Zn^{2+}, Zn^+)$. Also, as Ni^{2+} oxidizes Zn^+ then $E°(Ni^{2+}, Ni^+) > E°(Zn^{2+}, Zn^+)$. Inorganic ions like NO_3^-, MnO_4^-, IO_3^- and BrO_3^- are reduced also by e^-_{aq} with rate constants between 10^9 and 10^{10} M^{-1} s^{-1}. Chromate ion (CrO_4^{2-}) is reduced to what is presumed to be a Cr(V) species (Fig. 3.7).

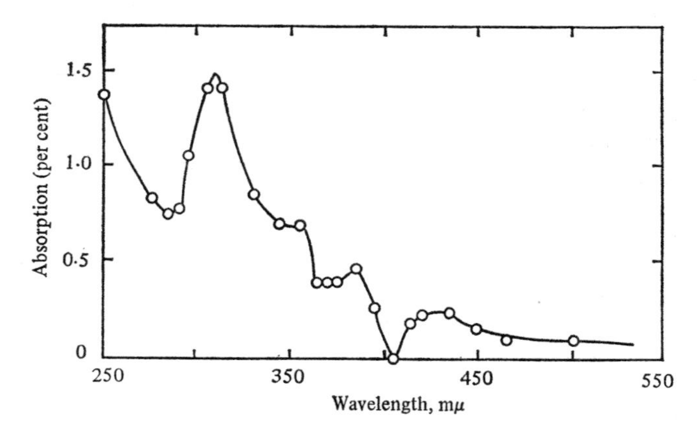

Fig. 3.7. Transient absorption due to a Cr(V) species, produced by an electron pulse on an aqueous solution of CrO_4^{2-}. From J. H. Baxendale, E. M. Fielden, and J. P. Keene, *Proc. Roy. Soc.*, 1965, **A286**, 320.

Ligands exert a pronounced influence on the reactivities of metal complexes toward e^-_{aq}.[85] The order of increasing efficiency of bridging ligands is $OH^- < CN^- < NH_3 < H_2O < F^- < Cl^- < I^-$, a sequence which parallels that observed for normal redox processes of the outer-sphere type. Thus the function of the ligand may be to act as an electron mediator, providing a route along which the electron may be channelled to the metal ion. An additional factor may arise from the effect of the ligand field on the energy levels of the *d* electrons. For example, cyanide ion exerts a strong field and gives rise to a large ligand-field splitting (Δ) so that, in the case of $[Fe(CN)_6]^{4-}$ reacting with e^-_{aq}, the added electron would have to enter a vacant high-energy e_g orbital. This would explain why the complex reacts only slowly with e^-_{aq}($k < 10^7$ M^{-1} s^{-1}).

References

1. C. F. Deck and A. C. Wahl, *J. Amer. Chem. Soc.*, 1954, **76**, 4054.
2. D. R. Stranks, *Disc. Faraday Soc.*, 1960, **29**, 73.
3. R. A. Marcus, *J. Phys. Chem.*, 1963, **67**, 853.
4. H. Taube, H. Myers, and R. L. Rich, *J. Amer. Chem. Soc.*, 1953, **75**, 4118.
5. A. Haim and W. K. Wilmarth, *J. Amer. Chem. Soc.*, 1961, **83**, 509.
6. J. H. Espenson, *Inorg. Chem.*, 1965, **4**, 1533.
7. J. C. Sullivan, *J. Amer. Chem. Soc.*, 1962, **84**, 4256; *Inorg. Chem.*, 1964, **3**, 315.
8. T. W. Newton and F. B. Baker, *Inorg. Chem.*, 1962, **1**, 368; G. Gordon, *Inorg. Chem.*, 1963, **2**, 1277.
9. T. W. Newton and F. B. Baker, *J. Phys. Chem.*, 1963, **67**, 1425.
10. J. H. Espenson, *Inorg. Chem.*, 1965, **4**, 1025; A. G. Sykes, *Chem. Commun.*, 1965, 442.

11(a). M. Ardon and R. A. Plane, *J. Amer. Chem. Soc.*, 1959, **81**, 3197.
11(b). R. W. Kolaczkowski and R. A. Plane, *Inorg. Chem.*, 1964, **3**, 322.
11(c). L. S. Hegedus and A. Haim, *Inorg. Chem.*, 1967, **6**, 664.

12. T. W. Newton and F. B. Baker, *Inorg. Chem.*, 1964, **3**, 569.
13. T. J. Conocchioli, E. J. Hamilton, and N. Sutin, *J. Amer. Chem. Soc.*, 1965, **87**, 926.
14. J. P. Candlin, J. Halpern, and S. Nakamura, *J. Amer. Chem. Soc.*, 1963, **85**, 2517.
15. J. Halpern and S. Nakamura, *J. Amer. Chem. Soc.*, 1965, **87**, 3002.
16. J. H. Espenson and J. P. Birk, *J. Amer. Chem. Soc.*, 1965, **87**, 3280.
17. A. Haim and N. Sutin, *J. Amer. Chem. Soc.*, 1966, **88**, 434.
18. A. Haim and N. Sutin, *J. Amer. Chem. Soc.*, 1965, **87**, 4210.
19. A. Haim and N. Sutin, *J. Amer. Chem. Soc.*, 1966, **88**, 5343.
20. T. J. Conocchioli, G. H. Nancollas, and N. Sutin, *J. Amer. Chem. Soc.*, 1964, **86**, 1453.
21. A. M. Zwickel and H. Taube, *J. Amer. Chem. Soc.*, 1961, **83**, 793.
22. J. P. Candlin, J. Halpern, and D. L. Trimm, *J. Amer. Chem. Soc.*, 1964, **86**, 1019.
23. J. H. Espenson, *Inorg. Chem.*, 1965, **4**, 121; D. L. Ball and E. L. King, *J. Amer. Chem. Soc.*, 1958, **80**, 1091.
24. A. Haim, *J. Amer. Chem. Soc.*, 1966, **88**, 2324.
25. D. E. Peters and R. T. M. Fraser, *J. Amer. Chem. Soc.*, 1965, **87**, 2758; R. T. M. Fraser, *Inorg. Chem.*, 1963, **2**, 954.
26. For reviews see H. Taube in *Advances in Chemistry Series* (ed. R. F. Gould), No. 49, American Chemical Society, 1965; N. Sutin, *Ann. Rev. Phys. Chem.*, 1966, **17**, 119.
27. See, for example, E. S. Gould, *J. Amer. Chem. Soc.*, 1966, **88**, 2983.
28. L. H. Sutcliffe and J. R. Weber, *Trans. Faraday Soc.*, 1956, **52**, 1225.
29. L. H. Sutcliffe and J. R. Weber, *J. Inorg. Nucl. Chem.*, 1960, **12**, 281.
30. J. H. Baxendale and C. F. Wells, *Trans. Faraday Soc.*, 1957, **53**, 800.
31. L. H. Sutcliffe and J. R. Weber, *Trans. Faraday Soc.*, 1959, **55**, 1892.
32. L. H. Sutcliffe and J. R. Weber, *Trans. Faraday Soc.*, 1961, **57**, 91.

33. L. E. Bennett and J. C. Sheppard, *J. Phys. Chem.*, 1962, **66**, 1275.
34. K. G. Ashurst and W. C. E. Higginson, *J. Chem. Soc.*, 1953, 3044.
35. A. G. Sykes, *J. Chem. Soc.*, 1961, 5549 (see also ref. (36)).
36. W. C. E. Higginson, D. R. Rosseinsky, J. B. Stead, and A. G. Sykes, *Disc. Faraday Soc.*, 1960, **29**, 49.
37. N. A. Daugherty, *J. Amer. Chem. Soc.*, 1965, **87**, 5026.
38. F. B. Baker, W. D. Brewer, and T. W. Newton, *Inorg. Chem.*, 1966, **5**, 1294.
39. F. R. Duke and C. E. Borchers, *J. Amer. Chem. Soc.*, 1953, **75**, 5186.
40. E. L. King and M. L. Pandow, *J. Amer. Chem. Soc.*, 1953, **75**, 3063.
41. M. K. Dorfman and J. W. Gryder, *Inorg. Chem.*, 1962, **1**, 799.
42. K. G. Ashurst and W. C. E. Higginson, *J. Chem. Soc.*, 1956, 343.
43. R. W. Dundon and J. W. Gryder, *Inorg. Chem.*, 1966, **5**, 986.
44. R. C. Thompson and J. C. Sullivan, *J. Amer. Chem. Soc.*, 1967, **89**, 1098.
45. D. R. Rosseinsky and W. C. E. Higginson, *J. Chem. Soc.*, 1960, 31.
46. W. H. McCurdy and G. G. Guilbault, *J. Phys. Chem.*, 1960, **64**, 1825.
47. D. R. Rosseinsky, *J. Chem. Soc.*, 1963, 1181.
48. R. H. Betts, *Canad. J. Chem.*, 1955, **33**, 1780.
49. S. W. Rabideau and R. J. Kline, *J. Phys. Chem.*, 1959, **63**, 1502.
50. S. W. Rabideau and R. J. Kline, *J. Phys. Chem.*, 1960, **64**, 193.
51. D. Benson and L. H. Sutcliffe, *Trans. Faraday Soc.*, 1960, **56**, 246.
52. D. Benson, P. J. Proll, L. H. Sutcliffe, and J. Walkley, *Disc. Faraday Soc.*, 1960, **29**, 60.
53. E. A. M. Wetton and W. C. E. Higginson, *J. Chem. Soc.*, 1965, 5890.
54. B. M. Gordon and A. C. Wahl, *J. Amer. Chem. Soc.*, 1958, **80**, 273.
55. A. C. Harkness and J. Halpern, *J. Amer. Chem. Soc.*, 1959, **81**, 3526.
56. See, for example, S. Gilks, T. E. Rogers, and G. M. Waind, *Trans. Faraday Soc.*, 1961, **57**, 1373.
57. F. A. Jones and E. S. Amis, *J. Inorg. Nucl. Chem.*, 1964, **26**, 1045.
58. A. M. Armstrong and J. Halpern, *Canad. J. Chem.*, 1957, **35**, 1020.
59. S. W. Rabideau and B. J. Masters, *J. Phys. Chem.*, 1961, **65**, 1256.
60. J. C. Sullivan, *J. Amer. Chem. Soc.*, 1965, **87**, 1495.
61. J. H. Espenson, *J. Amer. Chem. Soc.*, 1964, **86**, 5101.
62. N. A. Daugherty and T. W. Newton, *J. Phys. Chem.*, 1964, **68**, 612.
63. J. Y. Tong and E. L. King, *J. Amer. Chem. Soc.*, 1960, **82**, 3805.
64. J. B. Kirwin, P. J. Proll, and L. H. Sutcliffe, *Trans. Faraday Soc.*, 1964, **60**, 119.
65. D. H. Huchital, N. Sutin, and B. Warnqvist, *Inorg. Chem.*, 1967, **6**, 838.
66. W. C. E. Higginson and A. G. Sykes, *J. Chem. Soc.*, 1962, 2841.
67. T. W. Newton and F. B. Baker, *J. Phys. Chem.*, 1966, **70**, 1943.
68. J. C. Sheppard, *J. Phys. Chem.*, 1964, **68**, 1190.
69. S. W. Rabideau, *J. Phys. Chem.*, 1958, **62**, 414.
70. J. Halpern and J. G. Smith, *Canad. J. Chem.*, 1956, **34**, 1419.
71. P. George, *J. Chem. Soc.*, 1954, 4349.
72. E. Peters and J. Halpern, *J. Phys. Chem.*, 1955, **59**, 793; J. Halpern, E. R. Macgregor, and E. Peters, *J. Phys. Chem.*, 1956, **60**, 1455.

73. E. A. Hahn and E. Peters, *Canad. J. Chem.*, 1961, **39**, 162.
74. J. Halpern, G. Czapski, J. Jortner, and G. Stein, *Nature Lond.*, 1960, **186**, 629.
75. A. H. Webster and J. Halpern, *J. Phys. Chem.*, 1957, **61**, 1239.
76. A. H. Webster and J. Halpern, *Trans. Faraday Soc.*, 1957, **53**, 51.
77. G. J. Korinek and J. Halpern, *J. Phys. Chem.*, 1956, **60**, 285.
78. J. Kwiatek, I. L. Mador, and J. K. Seyler, in *Advances in Chemistry Series* (ed. R. F. Gould), No. 37, American Chemical Society, 1963.
79. A. C. Harkness and J. Halpern, *J. Amer. Chem. Soc.*, 1961, **83**, 1258.
80. E. J. Hart and J. W. Boag, *J. Amer. Chem. Soc.*, 1962, **84**, 4090.
81. L. M. Dorfman and M. S. Matheson, in *Progress in Reaction Kinetics* (ed. G. Porter), Vol. 3, p. 258, Pergamon, 1965.
82. J. K. Thomas, S. Gordon, and E. J. Hart, *J. Phys. Chem.*, 1964, **68**, 1524.
83. G. E. Adams, J. H. Baxendale, and J. W. Boag, *Proc. Chem. Soc.*, 1963, 241; J. H. Baxendale, E. M. Fielden, and J. P. Keene, *Proc. Chem. Soc.*, 1963, 242.
84. J. H. Baxendale, J. P. Keene and D. A. Stott, *Chem. Commun.*, 1966, 715.
85. M. Anbar and E. J. Hart, *J. Phys. Chem.*, 1965, **69**, 973.

Bibliography

H. Taube, Mechanisms of Redox Reactions of Simple Chemistry, in *Advances in Inorganic Chemistry and Radiochemistry* (ed. H. J. Emeléus and A. G. Sharpe), Vol. 1, p. 1, Academic Press, 1959.

J. Halpern, Mechanisms of Electron Transfer and Related Processes in Solution, in *Quarterly Reviews*, 1961, **15**, 207.

N. Sutin, The Kinetics of Inorganic Reactions in Solution, in *Annual Review of Physical Chemistry*, 1966, **17**, 119.

N. Sutin, Electron Exchange Reactions, in *Annual Review of Nuclear Science*, 1962, **12**, 285.

Oxidation-Reduction Processes in Ionizing Solvents, *Discussions of the Faraday Society*, No. 29, 1960.

R. A. Marcus, Chemical and Electrochemical Electron-Transfer Theory, in *Annual Review of Physical Chemistry*, 1964, **15**, 155.

C. B. Amphlett, Isotopic Exchange Reactions between Different Oxidation States in Aqueous Solution, in *Quarterly Reviews*, 1954, **8**, 219.

D. R. Stranks and R. G. Wilkins, Isotopic Tracer Investigations of Mechanism and Structure in Inorganic Chemistry, in *Chemical Reviews*, 1957, **57**, 743.

Solvated Electron, *Advances in Chemistry Series* (ed. R. F. Gould), No. 50, American Chemical Society, 1965.

J. Halpern, Catalysis by Coordination Compounds, in *Annual Review of Physical Chemistry*, 1965, **16**, 103.

I. Ruff, The Theory of Thermal Electron-transfer Reactions in Solution, in *Quarterly Reviews*, 1968, **22**, 199.

4. Reactions of oxoanions

The first part of this chapter is concerned with the reactions (both replacement and redox) of oxoanions of the halogens and sulphur. An important factor in the reactions of oxoanions is the breaking of the central atom–oxygen bond and, as a consequence, such reactions have features in common with the ligand replacement processes of metal complexes. However, a distinguishing feature of oxoanion reactions is their pronounced response to changes in acidity of the medium, since the labilizing effect of a proton is required in order that an oxygen can break free from its central atom.

In the second part of the chapter, consideration is given to the oxoanions of chromium and manganese. The material on the reactions of chromium(VI) supplements that in the previous chapter.

Exchange reactions of hypobromite and hypochlorite

The isotopic exchange of bromine between hypobromite and bromide ions has been studied in concentrated alkaline solution.[1] The rate of exchange is directly proportional to hypobromite and bromide concentrations and is inversely proportional to the hydroxide-ion concentration. This result suggests that the active species in the reaction is free hypobromous acid whose concentration is fixed by the equation

$$[HOBr] = \frac{K_W[BrO^-]}{K_A[OH^-]}$$

where K_A and K_W are, respectively, the acid dissociation constant of HOBr and the ionic product of water. The rate law

$$\text{rate} = k'[BrO^-][Br^-]/[OH^-]$$

is thus equivalent to

$$\text{rate} = k[HOBr][Br^-]$$

where $k = k'K_A/K_W$.

The reaction of hypobromite (and hypochlorite) with nitrite proceeds with almost complete transfer of the hypohalite oxygen to the reducing agent. This observation has afforded a means whereby the oxygen exchange of hypohalite with water can be followed.[2] Samples of solutions containing hypohalite and ^{18}O-enriched water were sampled at intervals and the ^{18}O-content of the hypohalite analysed indirectly by the addition of nitrite ions. Unreacted hypohalite was destroyed by peroxide, and nitrate precipitated with the reagent nitron; the nitron nitrate on evaporation to dryness yielded ammonium nitrate from which nitrous oxide was obtained for isotopic analysis by mass spectrometry. The more direct method of sampling the solvent was considered too inaccurate for dilute solutions of the solute. Under the conditions of the experiments the principal rate law for exchange was established as

$$\text{rate} = k_a'[XO^-]/[OH^-] = k_a[HOX]$$

The exchange is catalysed markedly by halide ion

$$\text{rate} = k_b'[XO^-][X^-]/[OH^-] = k_b[HOX][X^-]$$

It is seen that the derived rate law for the bromide-catalysed exchange of oxygen between hypobromite and solvent has a similar form to that of bromine exchange. The results of the two exchanges are given in Table 4.1. The fact that the two reactions have rate

Table 4.1

Comparison of the bromide-catalysed oxygen and bromine exchanges of hypobromous acid*

Reaction	$10^{-5}k$, M^{-1} min^{-1}, at 25°	Effect of Cl^-
Oxygen exchange	$1{\cdot}8 \pm 0{\cdot}1$	Positive, stronger than that of Br^-
Bromine exchange	$7{\cdot}6 \pm 1{\cdot}1$	Negative

* From ref. (1).

constants differing by a factor of four implies that the mechanism of interaction between HOBr and bromide ion is different in each case. Two mechanisms are indeed possible. Bromide may attack the bromine atom of the acid, displacing a hydroxyl group

$$\mathrm{HOBr + Br^- \rightleftharpoons \underset{\vdots \atop Br^-}{HOBr} \rightleftharpoons HO^- + \underset{| \atop Br}{Br}} \qquad (1)$$

This will result in oxygen exchange with the solvent simultaneous with bromine exchange. Alternatively, bromide may attack the oxygen atom displacing bromide ion from the acid

$$\mathrm{HOBr + Br^- \rightleftharpoons \underset{\vdots \atop Br^-}{HOBr} \rightleftharpoons \underset{| \atop Br}{HO} + Br^-} \qquad (2)$$

As a consequence bromine exchange will not result in oxygen transfer. It seems conceivable that bromine exchange operates by mechanism (2) and that exchange of oxygen in the presence of bromide occurs by mechanism (1). However, the rates do not differ sufficiently to allow such a discrimination to be made with complete confidence. A possible explanation for the catalytic effect of chloride on the rate of oxygen exchange is the formation of BrCl by

$$\mathrm{HOBr + Cl^- \rightleftharpoons \underset{\vdots \atop Cl^-}{HOBr} \rightleftharpoons HO^- + \underset{| \atop Cl}{Br}}$$

The chlorine exchange between hypochlorite and chloride ions, in the pH range 9·6–13·7, has been found to be first order in hypochlorite and chloride concentrations and second order in hydrogen-ion concentration

$$\text{rate} = k'[ClO^-][Cl^-][H^+]^2 = k[HOCl][Cl^-][H^+]$$

The chloride-catalysed oxygen exchange between HOCl and water is faster than the chlorine exchange. Also the former process is hydrogen-ion independent. It has been postulated that both processes proceed through a common intermediate $HOClCl^-$ which, in the case of chlorine exchange, is protonated and hydrated in the transition state as $H_2OClCl.H_2O$

$$\mathrm{HOCl.H_2O + H^+ \rightleftharpoons H_2{}^+OCl.H_2O}$$

$$\mathrm{H_2{}^+OCl.H_2O + \overset{*}{Cl}{}^- \rightleftharpoons H_2OCl\overset{*}{Cl}.H_2O}$$

$$\mathrm{H_2OCl\overset{*}{Cl}.H_2O \rightleftharpoons H_2O + Cl^- + \overset{*}{Cl}OH_2{}^+}$$

$$\mathrm{\overset{*}{Cl}OH_2{}^+ \rightleftharpoons \overset{*}{Cl}OH + H^+}$$

It is evident that the HOCl and HOBr systems are basically different in behaviour. This may be attributed to the relative strengths of the HO—X bond, the strong bond of HOCl necessitating the labilizing effect of an additional proton.

Further reactions of hypochlorite

Hypochlorite is unstable in alkaline solution, the rate of decomposition being proportional to the second power of the hypochlorite concentration.[3] Some oxygen is produced but chlorate and chloride are the main products; the latter has no specific effect on the rate. It is likely, therefore, that the reaction involves the steps

$$2ClO^- \rightarrow Cl^- + ClO_2^- \qquad \text{rate-determining}$$

$$ClO^- + ClO_2^- \rightarrow Cl^- + ClO_3^-$$

$$ClO^- \rightarrow Cl^- + \tfrac{1}{2}O_2$$

The individual rates are such that, at 40°, a solution of hypochlorite contains about 1 per cent chlorite. The reaction is catalysed heterogeneously by small quantities of cobalt, nickel, and copper (as metallic oxides). However, no effect on the rate of ClO_3^- production was observed, only an increase in the rate of oxygen evolution. A possible mechanism is

$$2MO + ClO^- \rightarrow M_2O_3 + Cl^- \qquad (1)$$

$$M_2O_3 + ClO^- \rightarrow M_2O_3.ClO^- \quad \text{(adsorbed)} \qquad (2)$$

$$M_2O_3.ClO^- \rightarrow 2MO + Cl^- + O_2 \qquad (3)$$

In the case of copper catalysis the rate is proportional to hypochlorite concentration and reaction (2) is rate-determining; for the catalysis by cobalt and nickel the catalyst surface is completely covered and (3) is the slow step, with the rate becoming zero-order in hypochlorite.

Nitrite is oxidized to nitrate by hypochlorite in alkaline solution.[4] Since the rate is inversely proportional to hydroxide-ion concentration, the slow step is indicated as

$$HOCl + NO_2^- \rightarrow Cl^- + H^+ + NO_3^-$$

with the pre-equilibrium

$$OCl^- + H_2O \rightleftharpoons HOCl + OH^-$$

A similar mechanism of attack on oxygen operates also in the reaction between hypochlorite and iodide ions, studied using a rapid-flow technique[5]

$$OCl^- + H_2O \underset{}{\overset{K_h}{\rightleftharpoons}} HOCl + OH^- \qquad \text{rapid equilibrium}$$

$$HOCl + I^- \xrightarrow{k_1} HOI + Cl^- \qquad \text{slow}$$

$$HOI + OH^- \rightleftharpoons IO^- + H_2O \qquad \text{rapid equilibrium}$$

with a rate law

$$-d[OCl^-]/dt = k_1 K_h [OCl^-][I^-]/[OH^-]$$

Some iodate and iodide are found as products since hypoiodite decomposes slowly

$$3OI^- \rightarrow IO_3^- + 2I^-$$

It is of interest to note an analogy between peroxides and hypochlorite in their reaction with bromide or iodide ions. It is highly probable that the rate-determining process in the peroxide reactions is the nucleophilic attack of the halide on a peroxo oxygen

$$R{-}O{-}O{-}H + Br^- \longrightarrow R{-}O{-}O(H)\cdots Br^- \xrightarrow{\text{slow}} RO^- + HOBr$$

followed by

$$HOBr + H_3O^+ + Br^- \xrightarrow{\text{fast}} Br_2 + 2H_2O$$

$$RO^- + H_3O^+ \xrightarrow{\text{fast}} ROH + H_2O$$

It is argued that the rates of different peroxides are related to the stability of the RO^- ion relative to ROH: if RO^- is stable then the products are formed more rapidly. A measure of the stability of RO^- is the pK_A value of ROH. Plots of log k versus pK_A of ROH are linear and the rate data for the reactions of hypochlorite with bromide and iodide fall nearly on these lines (assuming $RO^- = Cl^-$ and ROH = HCl). This comparison between the breaking of O—O and O—Cl bonds is justified on the grounds that the bonds have similar strengths.

Some reactions of hypochlorite are reactions of free hypochlorous acid (e.g., with NO_2^-, OCN^-, Br^-, I^-, $HC_2O_4^-$) whereas others seem to involve the OCl^- ion (e.g., with IO_3^-, ClO_2^-, SO_3^{2-}). There is, at present, no general explanation that can be offered for this behaviour.

Oxygen exchanges of bromate and chlorate with water

The exchange of oxygen between bromate ions and water in acid solution[6] conforms to a third-order rate law of the form

$$\text{rate} = k[H^+]^2[BrO_3^-]$$

This result is suggestive of a mechanism involving protonated species of bromate formed by the rapid equilibria

$$H^+ + BrO_3^- \rightleftharpoons HBrO_3$$

$$H^+ + HBrO_3 \rightleftharpoons H_2BrO_3^+$$

Two methods are then possible for the oxygen of bromate to exchange with water: either a unimolecular dissociation

$$H_2BrO_3^+ \rightarrow BrO_2^+ + H_2O$$

or a bimolecular nucleophilic displacement

$$H_2\overset{*}{O} + H_2BrO_3^+ \rightarrow H_2O + H_2Br\overset{*}{O}_3^+$$

The rate of exchange in heavy water is considerably faster than that in ordinary water, a result consistent with reaction via $H_2Br\overset{*}{O}_3^+$ since deuterium oxide is more acidic than protium oxide. The exchange of oxygen between chlorate and water[7] follows an identical form of rate law to that of BrO_3^-. The observation that chlorine dioxide has no influence on the rate rules out the possibility that ClO_2^+ is an intermediate in the exchange since chlorine dioxide exists in the equilibrium

$$2ClO_2 \rightleftharpoons ClO_2^+ + ClO_2^-$$

It is interesting that perchlorate ion shows remarkable inertness to oxygen exchange: in 9 M perchloric acid at 100° the half-life is estimated to be greater than 100 years. There is little doubt that the overall order of reactivity in chlorine oxoanions, given by $ClO_4^- < ClO_3^- < ClO^-$, is related essentially to the oxidation state of the central halogen atom, although chlorite would seem to behave anomalously in that it does not exchange oxygen with water in conditions under which it is stable.

Oxygen exchange of iodate and water

The isotopic exchange of oxygen between iodate ions and water originally eluded study since it proved difficult to separate iodate from aqueous solution before it had attained isotopic equilibrium with the solvent. However, by careful experimentation and propitious choice of conditions, interesting rate data have been obtained.[8] The reaction is found to be both acid and base catalysed, the rate of exchange passing through a minimum at pH 7·7 and following a first-order dependence both in hydrogen-ion and hydroxide-ion concentration, over the pH range 5–10·6 (Fig. 4.1). Added iodate ion has no detect-

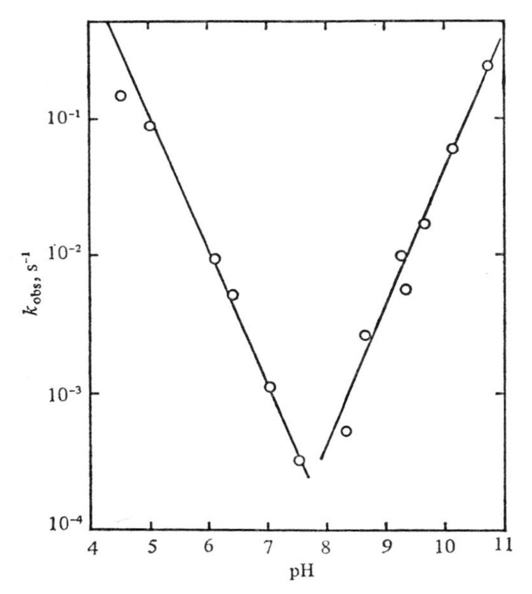

Fig. 4.1. The rate of exchange of oxygen between iodate and water as a function of pH. From M. Anbar and S. Guttmann, *J. Amer. Chem. Soc.*, 1961, **83**, 781.

able influence on the rate. The rate of exchange was drastically affected, however, by the presence of various ions and organic bases; this catalysis may explain why the reaction was originally considered too fast to measure. Catalytic coefficients (k_{acid} and k_{base}) were calculated for the conjugate acid–base pairs and are given in Table 4.2. It is seen that k_{acid} increases with increasing acid strength of the catalysts and that k_{base} follows a sequence of increasing basicity.

Iodate ion exists in solution in the form of various hydrated species, the simplest being $H_2IO_4^-$. The equilibrium of this ion with iodate

$$H_2IO_4^- \rightleftharpoons H_2O + IO_3^-$$

is probably the route through which oxygen exchange occurs with the solvent in acid solution. In more alkaline solutions $H_2IO_4^-$ dissociates according to

$$H_2IO_4^- \rightleftharpoons H^+ + HIO_4^{2-}$$

The decrease in rate of exchange with increasing alkalinity is due then to the increase in concentration of the HIO_4^{2-} ion which is incapable

Table 4.2

Catalytic coefficients (at 25°) of various species in the IO_3^-–H_2O exchange*

Base	k_{base}, M^{-1} s^{-1}	Acid	k_{acid}, M^{-1} s^{-1}
OH^-	1350	H_2O	10^{-4}
NH_3	270	NH_4^+	0·9
Pyridine	60	...	...
CN^-	45	HCN	0·9
$H_2BO_3^-$	7·5	H_3BO_3	0·6
Urea	5·4	...	...
Acetate	2·2	Acetic acid	12·3
F^-	0·09	...	...
H_2O	10^{-4}	H_3O^+	$3{\cdot}24 \times 10^4$
I^-	10^{-3}	...	...
IO_3^-	10^{-3}	...	...

* From ref. (8).

of undergoing reversible dehydration. At higher pH values a bimolecular nucleophilic substitution becomes assertive

$$\overset{*}{O}H^- + HIO_4^{2-} \rightleftharpoons OH^- + HI\overset{*}{O}_4^{2-}$$

and thus the rate of exchange begins to increase. The minimum in Fig. 4.1 is thus accounted for. General acid catalysis (by acetic and other acids) is probably due to the direct interaction of these acids with HIO_4^{2-}

$$HA + HIO_4^{2-} \rightleftharpoons A^- + H_2IO_4^- \rightleftharpoons A^- + H_2O + IO_3^-$$

General base catalysis may be due to the replacement of an OH^- ion by the base (B) in HIO_4^{2-}, resulting in the formation of a complex BIO_3^-

$$B + HIO_4^{2-} \rightleftharpoons BIO_3^- + OH^-$$

The existence of complexes of iodate with amines, pyridine, and fluoride has been substantiated and it seems likely that similar complexes can be formed with cyanide and borate.

Oxygen exchange of periodate and water

The oxygen exchange of periodate and water cannot be followed by the conventional tracer technique since the half-life at pH 4–7 is less than 5 seconds. Recently, however, the system has been studied successfully by oxygen-17 nuclear magnetic resonance.[9] In aqueous solutions of periodate the water ^{17}O line is broadened and no signal due to the ^{17}O of the periodate is observed. By investigating the effect of concentration and temperature on the line width, the kinetic features of the exchange were revealed. The species present in aqueous solutions of periodate are H_5IO_6, $H_4IO_6^-$, IO_4^-, and $H_3IO_6^{2-}$ which are involved in the following equilibria

$$H_5IO_6 \rightleftharpoons H^+ + H_4IO_6^- \qquad K_1 = 5{\cdot}1 \times 10^{-4}\ M$$

$$H_4IO_6^- \rightleftharpoons IO_4^- + 2H_2O \qquad K_2 = 40$$

$$H_4IO_6^- \rightleftharpoons H^+ + H_3IO_6^{2-} \qquad K_3 = 2{\cdot}0 \times 10^{-7}\ M$$

In acid solution undissociated H_5IO_6 is the dominant species. The exchange of oxygen between periodate and solvent in acid solution can be represented by

$$H_5IO_6 + H_2\overset{*}{O} \rightleftharpoons H_5IO_5\overset{*}{O} + H_2O$$

together with similar equations for the anionic species present. For oxoanions like iodate and periodate with large central atoms, the rate-controlling step is likely to be the elimination of water molecules from the anion rather than a nucleophilic displacement. This is indicated by the faster exchanges of iodate and periodate (compared with bromate and nitrate), and the fact that the periodate reaction is not catalysed by chloride (which is effective as a catalyst in the exchange of bromate and nitrate). In the case of periodate the slow step is the monomolecular dissociative process

$$H_5IO_6 \rightleftharpoons H_3IO_5 + H_2O$$

The analogous dehydration of $H_4IO_6^-$ has been studied by the temperature-jump method.[10]

Interesting comparisons can be made with the oxygen exchanges of the oxoacids of tellurium[11] and xenon[12] ($Te(OH)_6$ and $Xe(OH)_6$). Along with iodine, these are the largest non-metallic elements forming oxo-compounds and, as a consequence of their large radii, these elements form compounds of high coordination number in which the X—O bond possesses relatively little double-bond character and is relatively weak (in comparison with other smaller non-metallic oxo-compounds with low coordination numbers). It is plausible that the very rapid exchanges of telluric and xenic acids take place by a dissociative mechanism also. Perxenic acid ($Xe(OH)_8$) may be expected to exchange still more rapidly. For isoelectronic oxo-compounds of small central atoms in a low coordination state, the rate of exchange decreases as the charge of the central atom increases (e.g., $H_2Si^{IV}O_4^{2-} > HP^{V}O_4^{2-} > S^{VI}O_4^{2-} > Cl^{VII}O_4^{-}$). The converse is found on comparing the rates of periodic acid and telluric acid ($H_6Te^{VI}O_6 < H_5I^{VII}O_6$).

The chlorite–iodide reaction

Chlorite oxidizes iodide quantitatively according to

$$ClO_2^- + 4H^+ + 4I^- \rightarrow Cl^- + 2H_2O + 2I_2$$

The reaction has been followed in detail[13] by a spectrophotometric method over the pH range 4–8, and the existence of an autocatalytic term in the rate law is established

$$\mathrm{d}([I_2] + [I_3^-])/\mathrm{d}t = -2\mathrm{d}[ClO_2^-]/\mathrm{d}t = k'[ClO_2^-][I^-][H^+] + k''[ClO_2^-][I_2]/[I^-]$$

The second term predominates under conditions where significant concentrations of iodine are present; when iodine is present initially the reaction proceeds almost exclusively by this route. The individual rate constants at 25° are $k' = 9{\cdot}2 \times 10^2\ \mathrm{M^{-2}\,s^{-1}}$ and $k'' = 5{\cdot}1 \times 10^{-3}\ \mathrm{s^{-1}}$. On the addition of excess chlorite to iodide an initial slow increase of iodine occurs, followed, as the autocatalytic path takes over, by a rapid exponential increase. When 60–80 per cent of the theoretical amount of iodine has been produced there is a sudden and spectacular removal of iodine (Fig. 4.2); at this point the system gives a negative test for iodide. In the latter stages of reaction it appears, therefore, that the product of reaction is iodate, the processes $I^- \rightarrow I_2$ and $I_2 \rightarrow IO_3^-$ proceeding simultaneously. A similar event takes place when iodine is present initially except that the induction period is absent.

The first term in the rate law can be accounted for by the sequence

$$H^+ + ClO_2^- \underset{}{\overset{K_1}{\rightleftharpoons}} HClO_2 \qquad \text{rapid equilibrium, } K_1 = 10^2\ M^{-1}$$

$$HClO_2 + I^- \xrightarrow{k_1} HOCl + IO^- \quad \text{rate-determining}$$

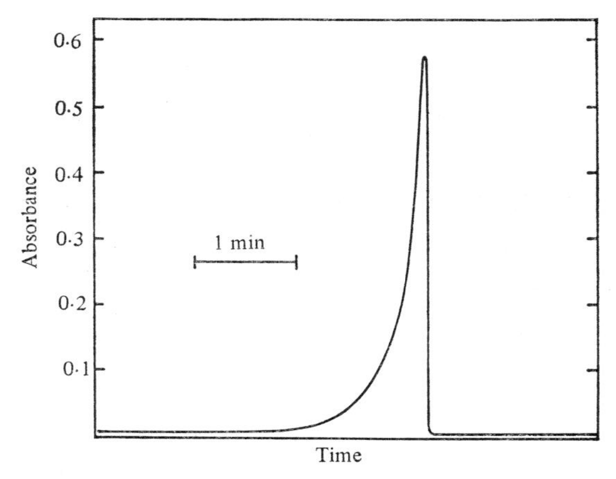

Fig. 4.2. Absorbance–time curve for iodine in the chlorite + iodide reaction: 467 mμ, cell length 10 cm, $[I^-] = 2{\cdot}7 \times 10^{-4}$ M, $[ClO_2^-] = 2{\cdot}5 \times 10^{-3}$ M, pH 5·8, 25°. From D. M. Kern and C. H. Kim, *J. Amer. Chem. Soc.*, 1965, **87**, 5309.

followed by rapid reduction of the products by excess iodide. The derived rate law is

$$d([I_2] + [I_3^-])/dt = 2k_1[HClO_2][I^-] = 2k_1K_1[ClO_2^-][I^-][H^+]$$

where the observed rate constant, k', is given by $k' = 2k_1K_1$, allowing the calculation of k_1 as $4{\cdot}6\ M^{-1}\ s^{-1}$ at 25°. The second term of the rate law corresponds to a mechanism

$$I_2 + H_2O \overset{K_2}{\rightleftharpoons} I^- + HOI + H^+ \quad \text{rapid equilibrium,}$$

$$K_2 = 4 \times 10^{-13}\ M^2$$

$$HOI + HClO_2 \xrightarrow{k_2} HIO_2 + HOCl \quad \text{rate-determining}$$

Under conditions of excess iodide, further reaction occurs to give a net yield of two iodine molecules per chlorite; at very low concentrations of iodide, HIO_2 is oxidized instead to iodate. The rate law for this route follows as

$$d([I_2] + [I_3^-])/dt = 2k_2[HOI][HClO_2] = 2k_2 K_1 K_2[ClO_2^-][I_2]/[I^-]$$

where $k'' = 2k_2 K_1 K_2$, from which $k_2 = 6 \times 10^7\ M^{-1}\ s^{-1}$ at 25°.

The reaction between iodate and iodide

The well-known analytical reaction, named after Andrews, between iodide and iodate in strongly acid solution

$$5I^- + IO_3^- + 6H^+ \rightleftharpoons 3I_2 + 3H_2O \qquad K \sim 10^{47} \qquad (1)$$

has been discussed [14] in terms of a mechanism using species which participate also in the disproportionation of hypoiodite. The scheme adopted, based upon the ionization of iodic acid, includes IO_2^+, I^+, and IO^+ as reactive entities, and can be expressed comprehensively as

$$HIO_3 \rightleftharpoons IO_2^+ + OH^-$$

$$IO_2^+ + 2I^- \rightarrow I^+ + 2IO^-$$

$$IO_2^+ + I^- \rightarrow IO^+ + IO^-$$

$$IO^+ + I^- \rightarrow I^+ + IO^-$$

$$IO^- + H^+ \rightarrow HOI$$

$$HOI + H^+ \rightarrow H_2OI^+$$

$$H_2OI^+ + I^- \rightarrow I_2 + H_2O$$

The exchange reaction between iodate and iodine is described by the very complex rate law

$$\text{rate} = k'[I_2]^{3/5}[H^+]^{9/5}[IO_3^-]^{9/5}$$

and is considered to occur through the iodate–iodide equilibrium (1).[15] Substitution of the concentration of free iodine given by the equilibrium expression into the rate law yields the sixth-order equation

$$\text{rate} = k''[I^-][H^+]^3[IO_3^-]^2$$

The periodate–iodide reaction

Periodate reacts with iodide ions in weakly acidic or neutral media according to

$$IO_4^- + 2I^- + 2H^+ \rightarrow IO_3^- + I_2 + H_2O$$

At low iodide-ion concentrations the subsequent reaction represented by

$$IO_3^- + 5I^- + 6H^+ \rightarrow 3I_2 + 3H_2O$$

is so slow that, under these conditions of low acidity, it may be disregarded. The rate of the first reaction is given by

$$\text{rate} = k_1[I^-][IO_4^-] + k_2[I^-][IO_4^-][H^+]$$

On the basis of salt effects it was concluded [16] that a transfer of an oxygen atom (or two OH groups) takes place from the periodate to the iodide to form hypoiodite. The latter then oxidizes a further iodide ion to iodine. The transfer of oxygen entails complete rearrangement of the octahedral structure of periodate into that of iodate which is pyramidal.

Reactions of halogen oxoanions with sulphite

In the reactions between halogen oxoanions and sulphite in an oxygen-labile solvent like water, water or the oxidizing agent may be the source of oxygen which is added to the reducing agent. Such systems have been investigated using ^{18}O as a tracer.[17] Reactions of ClO_3^-, ClO_2^-, and BrO_3^- with sulphite are sufficiently rapid, compared with their exchanges with water, to make the tracer method applicable without serious complications. In the case of chlorate, transfer of oxygen atoms occurs from the oxidant to the sulphite. However, the transfer is less than the maximum of three per ClO_3^- (2·3 in 0·1 M acid)

$$Cl\overset{*}{O}_3^- + 3H_2SO_3 \rightarrow Cl^- + 3H_2S\overset{*}{O}_4$$

Chlorite and sulphite react very rapidly, the reaction being essentially

$$Cl\overset{*}{O}_2^- + 2H_2SO_3 \rightarrow Cl^- + 2H_2S\overset{*}{O}_4$$

but a side reaction giving chlorate occurs and, again, less than the maximum transfer of oxygen atoms is found. The reaction of bromate and sulphite takes place rapidly and cleanly with bromide as the only halogen-containing product but again with incomplete transfer of oxygen.

Oxygen-atom transfer of chlorate and chlorite can be explained on the basis that reduction with sulphite proceeds step-wise by $ClO_3^- \rightarrow ClO_2^-$ and $ClO_2^- \rightarrow ClO^-$. The incomplete transfer of oxygen is due to the last stage of reduction in which hypochlorite is converted to chloride. It is apparent from the results that the path in which the oxygen added to the sulphate is derived from the solvent (forming a complex with a S—Cl bond) can make only a small contribution

$$2H^+ + SO_3^{2-} + ClO_3^- \rightarrow [O_3S\text{—}ClO_2^-] + H_2O \rightarrow SO_4^{2-} + 2H^+ + ClO_2^-$$

Two possibilities, which cannot be distinguished, remain: firstly, a direct replacement of ClO_2^- by SO_3^{2-} on oxygen

$$SO_3^{2-} + Cl\overset{*}{O}_3^- \xrightarrow{H^+} SO_3\overset{*}{O}^{2-} + Cl\overset{*}{O}_2^-$$

Secondly, a S—O—Cl bonded intermediate is formed by an esterification-type process

$$2H^+ + SO_3^{2-} + Cl\overset{*}{O}_3^- \rightarrow [O_2S\text{—}\overset{*}{O}\text{—}Cl\overset{*}{O}_2^-] + H_2O \rightarrow SO_3\overset{*}{O}^{2-} + 2H^+ + Cl\overset{*}{O}_2^-$$

In forming the S—O—Cl bond it is likely that the S—O bond is broken rather than O—Cl since sulphite exchanges with water much more rapidly than chlorate. It would seem that the disparity noted above, that less than the maximum number of oxygen atoms are transferred, is due to competition of

$$Cl\overset{*}{O}^- + SO_3^{2-} \rightarrow Cl^- + SO_3\overset{*}{O}^{2-}$$

with

$$Cl\overset{*}{O}^- + SO_3^{2-} + H^+ \rightarrow ClSO_3^- + \overset{*}{O}H^-$$

$$ClSO_3^- + H_2O \rightarrow SO_4^{2-} + 2H^+ + Cl^-$$

in which Cl^+ is transferred to form an intermediate which then undergoes hydrolysis.

Attention has been drawn to a correlation between the reactivity of halogenates and the π-bond character of the halogen-oxygen bonds.[18] The order of decreasing reactivity (as indicated by the comparative rates of oxygen exchange of the halogenate with water, and by the reaction of halogenate with sulphite) is

$$IO_3^-, IO_4^- > BrO_3^- > ClO_3^- > ClO_4^-$$

paralleling a similar sequence of increasing percentage π-bond character. Increase in π-bond character implies a greater utilization of the oxygen *p*-orbitals, thus rendering them less available for intermolecular bonding with solvent molecules. In this respect iodate and periodate, for which the I—O bond has little π-character, exchange oxygen with water through hydrated species ($H_2IO_4^-$ and H_5IO_6) whereas there is no such evidence in the case of bromate and chlorate. Also, as the experimental values for bond length are in the order IO_3^-, $IO_4^- > BrO_3^- > ClO_3^-$, ClO_4^-, there is some justification for the order of reactivity in terms of the relative ease of breaking of the X—O bond.

Reaction of sulphite with hydrogen peroxide

The rapid reaction between hydrogen peroxide and sulphite, represented by

$$H_2O_2 + H_2SO_3 \rightarrow H_2O + H_2SO_4$$

has been studied[19] using ^{18}O-enriched peroxide (prepared by passing $H_2{}^{18}O$ as a vapour at low pressure through a glow discharge and freezing out the effluent gas). No complications arose from the exchange of peroxide and water since, under the conditions employed, this is very slow. The results obtained are somewhat surprising. In acid solution (pH 5 and stronger) each molecule of peroxide which reacts transfers very nearly two atoms of oxygen to the sulphite; the overall change, however, requires only the transfer of a single oxygen atom. The explanation of this apparent anomaly is that peroxosulphurous acid is formed as an intermediate by the condensation

$$\mathrm{HO{-}\overset{\overset{\displaystyle O}{|}}{S}{-}OH + H{-}\overset{*}{O}{-}\overset{*}{O}{-}H \rightarrow H{-}O{-}\overset{\overset{\displaystyle O}{|}}{S}{-}\overset{*}{O}{-}\overset{*}{O}{-}H + H_2O}$$

Formation of doubly-labelled sulphate then occurs by an intramolecular rearrangement of the ester. Peroxosulphurous acid is analogous to peroxonitrous acid, a known participant in the peroxide–nitrous acid system.

Nucleophilic displacements on sulphur

The decomposition of thiosulphate in acid solution has been studied extensively over a period of some 80 years. The reaction is a complex one with a diversity of products consisting of several forms of elemental sulphur (S_6, S_8, and S_x) in the colloidal state, sulphur dioxide, hydrogen sulphide, and polythionic acids ($H_2S_xO_6$). La Mer and co-workers[20], from a detailed examination of the kinetics, have shown that the rate of appearance of colloidal sulphur (measured spectrophotometrically at 400 mμ) is given by

$$\text{rate} = k_1[S_2O_3^{2-}]^{3/2}[H^+]^{1/2}$$

and the rate of formation of sulphur dioxide by

$$\text{rate} = k_2[S_2O_3^{2-}]^2[H^+]$$

The kinetics are best interpreted by a scheme[21]

$$S_2O_3^{2-} + H^+ \rightleftharpoons HS_2O_3^-$$

$$HS_2O_3^- + S_2O_3^{2-} \rightarrow HS_3O_3^- + SO_3^{2-}$$

$$HS_3O_3^- + S_2O_3^{2-} \rightarrow HS_4O_3^- + SO_3^{2-}$$

$$HS_xO_3^- + S_2O_3^{2-} \rightarrow HS_{x+1}O_3^- + SO_3^{2-}$$

$$HS_8O_3^- + S_2O_3^{2-} \rightarrow HS_9O_3^- + SO_3^{2-}$$

the ions $HS_9O_3^-$ and $HS_7O_3^-$ being the precursors of sulphur in the forms S_8 and S_6

$$HS_9O_3^- \rightarrow S_8 + HSO_3^-$$

$$HS_7O_3^- \rightarrow S_6 + HSO_3^-$$

These reactions are viewed as nucleophilic displacements on sulphur, $S_2O_3^{2-}$ displacing the SO_3^{2-} group from the end of the chain:

$$H{-}S{-}\overset{\otimes}{S}O_3^- + S{-}SO_3^{2-} \rightarrow H{-}S{-}S{-}SO_3^- + \overset{\otimes}{S}O_3^{2-}$$

The formation of S_8 would seem to take place by an intramolecular rearrangement of the sulphopolysulphanide, $HS_9O_3^-$

```
  S--SH               S—S⁻                  S—S
S      S-SO₃⁻       S      S-SO₃H         S      S + HSO₃⁻
|      |      ⇌     |      |       ⟶      |      |
S      S            S      S              S      S
  S—S                 S—S                   S—S
```

Hexatomic sulphur, formed by a similar rearrangement of $HS_7O_3^-$, occurs in low yield due to the comparative instability of the S_6 ring. Polythionates and hydrogen sulphide are formed by thiophilic displacements of the type

$$\begin{array}{c}SH\\|\\S\\|\\SO_3^-\end{array} \rightleftharpoons \begin{array}{c}S^-\\|\\S\\|\\SO_3H\end{array} + \begin{array}{c}\overset{\otimes}{S}\text{—}SH\\|\\SO_3^-\end{array} \rightarrow \begin{array}{c}S\text{—}S\\|\quad|\\S\quad SO_3^-\\|\\SO_3H\end{array} + \overset{\otimes}{S}H^-$$

The reaction between tetrathionate and cyanide affords a further example of a nucleophilic displacement at the sulphur atom.[22] The products of the quantitative reaction are sulphate, thiosulphate, and thiocyanate

$$CN^- + S_4O_6^{2-} + H_2O \xrightarrow{k} SO_4^{2-} + S_2O_3^{2-} + SCN^- + 2H^+$$

A complication arises in that thiosulphate reacts further with cyanide ion forming sulphite and thiocyanate; however, this process is very slow compared with the main reaction. The rate data fit the simple expression

$$-d[S_4O_6^{2-}]/dt = k[S_4O_6^{2-}][CN^-]$$

Furthermore, when labelled tetrathionate (labelled with ^{35}S in the dithio group) reacts with cyanide the sulphate formed is inactive, whilst the thiocyanate is active and the first thiosulphur atom of the thiosulphate is active. These observations call for a mechanism involving, firstly, a nucleophilic attack of the cyanide ion on the sulphur atom displacing thiosulphate

$$CN^- + {}^-O_3S\text{—}S\text{—}S\text{—}SO_3^- \rightarrow {}^-O_3S\text{—}SCN + S\text{—}SO_3^{2-}$$

followed by the hydrolysis

$${}^-O_3S\text{—}SCN + 2OH^- \rightarrow SO_4^{2-} + SCN^- + H_2O$$

A low activation energy (in this case, of 11 kcal $mole^{-1}$) appears to be characteristic of displacement reactions on sulphur.

The oxidation of S(IV) by Cr(VI)

The oxidation of sulphur(IV) by chromium(VI) and manganese(VII) yields dithionate, in addition to sulphate, as the products of reaction. Iodine (as I_3^-), however, produces no dithionate. This is to be expected since the oxidation potential of the I^-/I_3^- couple is too low to permit the formation of dithionate, enabling sulphur(IV) to be oxidized completely to sulphur(VI). The stoichiometry of the Cr(VI)+ S(IV) system varies over a Cr(VI)/S(IV) mole ratio of 1:2 to 2:3 as the initial concentrations are altered in the range $[Cr(VI)]_0/[S(IV)]_0$ of 0·12 to 1·4. This variable stoichiometry can be explained on the basis of a competition between the two net reactions

$$2HCrO_4^- + 4HSO_3^- + 6H^+ \rightarrow 2Cr^{3+} + 2SO_4^{2-} + S_2O_6^{2-} + 6H_2O \quad (1)$$

$$2HCrO_4^- + 3HSO_3^- + 5H^+ \rightarrow 2Cr^{3+} + 3SO_4^{2-} + 5H_2O \quad (2)$$

A rate law was obtained for reaction (1) by an examination of the kinetics[23] under conditions of excess S(IV) (where the stoichiometry approaches 1:2) and over the pH range 4·2–5·0 (where the reactants are identifiable as the species $HCrO_4^-$ and HSO_3^-)

$$\frac{-d[Cr(VI)]}{dt} = \frac{k[Cr(VI)][S(IV)]^2[H^+]}{1+K[S(IV)]}$$

$HCrO_4^-$ is known to condense with itself and anions like HSO_4^- and $H_2PO_4^-$ to form $Cr_2O_7^{2-}$, $CrSO_7^{2-}$, and $HCrPO_7^{2-}$, respectively. By analogy it is assumed that $HCrO_4^-$ undergoes condensation with HSO_3^- forming $CrSO_6^{2-}$

$$HCrO_4^- + HSO_3^- \underset{}{\overset{K_1}{\rightleftharpoons}} CrSO_6^{2-} + H_2O$$

This pre-equilibrium accounts for the denominator term in the rate law. On the basis of this supposition a mechanism, consistent with the experimental observations, was developed

$$HSO_3^- + H^+ \overset{K_2}{\rightleftharpoons} SO_2 + H_2O$$

$$SO_2 + CrSO_6^{2-} \xrightarrow{k_1} \left[\begin{array}{c} O \\ | \\ O_2SOCrOSO_2 \\ | \\ O \end{array}\right]^{2-}$$

rate-determining

$$[CrO_2(SO_3)_2]^{2-} + 4H_2O + 2H^+ \rightarrow [SO_4Cr(H_2O)_5]^+ + SO_3^-$$

$$2SO_3^- \rightarrow S_2O_6^{2-}$$

These five steps explain the observed stoichiometry under conditions where reaction (1) is predominant, and generate an expression of the same form as the empirical rate equation

$$\frac{-\mathrm{d}[Cr(VI)]}{\mathrm{d}t} = \frac{k_1 K_1 K_2 [HCrO_4^-][HSO_3^-]^2[H^+]}{1 + K_1[HSO_3^-]}$$

The activated complex formed in the rate-determining step can decompose, assisted by solvent and protons, to produce a Cr(III)–sulphate complex and a dithionate radical-ion from which dithionate is formed. Thus the limiting stoichiometry of Cr(VI)/S(IV) = 1:2 is obtained, where half of the sulphur is oxidized to $S_2O_6^{2-}$ and half to SO_4^{2-}.

Modifications have to be made to this mechanism to account for the complete conversion of S(IV) to S(VI) under conditions of excess Cr(VI). The following additional reactions are postulated:

$$SO_3^- + HCrO_4^- \rightarrow SO_4^{2-} + Cr(V)$$

$$H^+ + HCrO_4^- + CrSO_6^{2-} \rightarrow O_3CrO\overset{\displaystyle O}{\underset{\displaystyle O}{\overset{|}{\underset{|}{Cr}}}}OSO_2^{2-} + H_2O$$

$$O_3CrO\overset{\displaystyle O}{\underset{\displaystyle O}{\overset{|}{\underset{|}{Cr}}}}OSO_2^{2-} \rightarrow 2Cr(V) + SO_4^{2-}$$

$$Cr(V) + S(IV) \rightarrow Cr(III) + S(VI)$$

The chromium(VI)–chromium(III) exchange

The exchange of chromium(III) and chromium(VI) in acid solution is very slow: at a total chromium concentration of 0·02 M less than 43 per cent exchange occurs in 1200 hours at 45°. This inertness is to be expected since there is a difference of three in the oxidation states of the two species. Furthermore, exchange necessitates a change in coordination number from 6, in the case of the aquochromium(III) ion, to 4, for chromium(VI) as chromate or dichromate. The kinetics of the reaction have been studied at 95° [24] where the rate is governed by the expression

$$\text{rate} = [Cr(OH_2)_6^{3+}]^{4/3}[H_2CrO_4]^{2/3}\{k[H^+]^{-2} + k'\}$$

The form of the hydrogen-ion independent term suggests that the transition state for the exchange is composed of two chromium atoms ($4/3 + 2/3 = 2$) with an average oxidation number of +4 for each atom ($4/3 \times (+3) + 2/3 \times (+6) = 2 \times (+4)$). From these, and other considerations, the rate-determining step is identified as a reaction between chromium(III) and chromium(V)

$$\overset{*}{\mathrm{Cr}}(\mathrm{III}) + \mathrm{Cr}(\mathrm{V}) \underset{}{\overset{\text{slow}}{\rightleftharpoons}} \overset{*}{\mathrm{Cr}}(\mathrm{V}) + \mathrm{Cr}(\mathrm{III})$$

which may occur alternatively as two separate 'one-electron transfers' via Cr(IV). Initially an equilibrium concentration of Cr(V) (proportional to $[Cr(III)]^{1/3}[Cr(VI)]^{2/3}$) is established by interaction of chromium(III) and chromium(VI)

$$\overset{*}{\mathrm{Cr}}(\mathrm{III}) + \mathrm{Cr}(\mathrm{VI}) \overset{\text{fast}}{\rightleftharpoons} \overset{*}{\mathrm{Cr}}(\mathrm{IV}) + \mathrm{Cr}(\mathrm{V})$$

the feasibility of which depends upon the similarities in coordination between Cr(VI) and Cr(V), and also Cr(III) and Cr(IV). In this respect, it is pertinent that the analogous equilibria of neptunium and plutonium are set up rapidly.

Completion of the overall process entails a rapid exchange

$$\overset{*}{\mathrm{Cr}}(\mathrm{V}) + \mathrm{Cr}(\mathrm{VI}) \overset{\text{fast}}{\rightleftharpoons} \overset{*}{\mathrm{Cr}}(\mathrm{VI}) + \mathrm{Cr}(\mathrm{V})$$

between chromium(V) and chromium(VI). Significantly, manganese(VI) and manganese(VII), which are isoelectronic with chromium(V) and chromium(VI) and are both tetrahedrally coordinated, undergo very fast exchange (see p. 182). The value obtained for the constant k' in the rate law is $1{\cdot}4 \times 10^{-5}$ M^{-1} s^{-1} at $\mu = 0{\cdot}92$ M.

The pH-dependence on the rate is indicative of the joint participation of the two monomeric chromium(VI) species, $HCrO_4^-$ and CrO_4^{2-}, the latter being present in appreciable concentrations ($k = 6{\cdot}6 \times 10^{-7}$ M^{-1} s^{-1}). Under conditions of high chromium(VI) concentrations there is a substantial proportion of dimeric species ($Cr_2O_7^{2-}$, and probably also of Cr(V)), but their particular role in the mechanism is impossible to disentangle since the reaction orders are modified but slightly on including them in the scheme. A further complication, precluding a full analysis of the system, is the indisputable existence of oxygen-bridged chromium(III)–chromium(VI) species, $CrCrO_4^+$ and $CrCrO_4H^{2+}$, in acid solution.[25]

Induced oxidations of Cr(VI)

The oxidation of iodide ion by Cr(VI)

$$2HCrO_4^- + 6I^- + 14H^+ \xrightarrow[\text{slow}]{\text{very}} 2Cr(III) + 3I_2 + 8H_2O$$

in dilute acid solution (0·001 N) and at low concentrations, is very slow. Also, under these conditions, the reaction between iron(II) and Cr(VI) is rapid

$$HCrO_4^- + 3Fe(II) + 7H^+ \xrightarrow{\text{rapid}} Cr(III) + 3Fe(III) + 4H_2O$$

whereas that between Fe(III) and iodide is very slow

$$2Fe(III) + 2I^- \xrightarrow[\text{slow}]{\text{very}} 2Fe(II) + I_2$$

However, in dilute acid solution, iodide is rapidly oxidized to iodine by Cr(VI) in the presence of small amounts of Fe(II), a reaction which can be represented accurately by the stoichiometric equation

$$HCrO_4^- + Fe(II) + 2I^- + 7H^+ \xrightarrow{\text{rapid}} Cr(III) + Fe(III) + I_2 + 4H_2O$$

or more simply by

$$Cr(VI) + Fe(II) + 2I^- \xrightarrow{\text{rapid}} Cr(III) + Fe(III) + I_2 \qquad (1)$$

Iron(II) is said to *induce* the reaction between Cr(VI) and iodide ion. Iron(II) is the *inductor*, chromium(VI) is the *actor*, and iodide ion is the *acceptor*. The reaction between the actor and the acceptor is referred to as the induced reaction, and that between actor and inductor as the primary reaction. It is evident that there is a fundamental difference between an induced reaction and a catalysed reaction since the inductor (Fe(II) in this case) is consumed in the primary reaction and is not regenerated. Another familiar example to which similar arguments apply, occurs in the oxidation of manganese(II) by chromium(VI) induced by arsenite

$$2HCrO_4^- + 2H_3AsO_3 + Mn(II) + 6H^+ \xrightarrow{\text{rapid}} 2Cr(III) + 2H_3AsO_4 + MnO_2 + 4H_2O$$

that is

$$2Cr(VI) + 2As(III) + Mn(II) \xrightarrow{\text{rapid}} 2Cr(III) + 2As(V) + Mn(IV) \qquad (2)$$

Here one equivalent of Mn(II) is oxidized for every two equivalents of As(III) oxidized. In the case of reaction (1), two equivalents of iodide are oxidized along with one equivalent of Fe(II).

These results are usually expressed in terms of an *induction factor* which is defined[26] as the ratio of the number of equivalents of the acceptor oxidized to the number of equivalents of inductor oxidized. The induction factors for reactions (2) and (1), respectively, are 0·5 and 2. Values of 0·5 or 2 are found for most induced chromium(VI) oxidations (see Table 4.3). It should be noted, however, that the factors quoted in Table 4.3 are limiting values obtaining only under conditions of high concentrations of acceptor relative to inductor.

Table 4.3

Some induced oxidations of Cr(VI)*

Inductor	Acceptor	Induction factor
Fe(II)	I^-	2
H_3AsO_3	I^-	2
$VOSO_4$	I^-	2
$(VO)_2SO_4$	I^-	2
VSO_4	I^-	2
Ti(III)	I^-	2
$UOSO_4$	I^-	0·5–0·7
H_3AsO_3	Mn(II)	0·5
Fe(II)	Br^-	unknown
$VOSO_4$	Br^-	unknown
Ti(III)	Br^-	unknown

* From ref. (26).

In other words, the overall change from Cr(VI) to Cr(III), a three-equivalent process, results in the oxidation of either one or two equivalents of inductor. Direct indication is thus given of the participation of either Cr(V) or Cr(IV) as transient intermediates in the reactions.

In the case of the chromium(VI)+iodide reaction induced by iron(II), Cr(V) is produced in the primary step by the one-equivalent process

$$\mathrm{Cr(VI) + Fe(II) \rightleftharpoons Cr(V) + Fe(III)} \quad (3)$$

followed by

$$\mathrm{Cr(V) + I^- \rightarrow Cr(III) + IO^-}$$

$$\mathrm{IO^- + I^- + 2H^+ \rightarrow I_2 + H_2O}$$

The primary reaction has been examined in some detail:[27] in the absence of iodide the steps subsequent to (3) are

$$\mathrm{Cr(V) + Fe(II)} \xrightarrow{\text{slow}} \mathrm{Cr(IV) + Fe(III)} \qquad \text{rate-determining}$$

$$\mathrm{Cr(IV) + Fe(II)} \xrightarrow{\text{fast}} \mathrm{Cr(III) + Fe(III)}$$

in accord with the kinetic expression

$$\frac{-\mathrm{d}[Fe(II)]}{\mathrm{d}t} = \frac{k[HCrO_4^-][Fe(II)]^2[H^+]^3}{[Fe(III)]}$$

In the presence of moderate concentrations of iodide ion, Fe(II) and iodide compete for the pentavalent chromium and the induction factor is less than 2, a limiting value which is attained only at high concentrations of iodide ion.

The reaction between chromium(VI) and manganese(II), induced by arsenite, can be reconciled in terms of a two-equivalent primary step

$$\mathrm{Cr(VI) + As(III) \rightleftharpoons Cr(IV) + As(V)} \tag{4}$$

producing Cr(IV) as an intermediate valency state. Cr(IV) then reacts further by

$$\mathrm{Cr(IV) + Mn(II) \rightarrow Cr(III) + Mn(III)}$$

yielding Mn(IV) as a product of the reaction by the disproportionation

$$\mathrm{2Mn(III) \rightleftharpoons Mn(II) + Mn(IV)}$$

The oxidation of As(III) by Cr(VI)

$$\mathrm{3As(III) + 2Cr(VI) \rightarrow 3As(V) + 2Cr(III)}$$

has been studied, in the absence of Mn(II), in 0·2 M acetic acid–acetate buffers (pH 4·6) at an ionic strength of 1·5 M.[28] Under conditions of high As(III) and low Cr(VI) the rate deviates from second-order behaviour and is given by

$$\frac{-\mathrm{d}[Cr(VI)]}{\mathrm{d}t} = \frac{kK[As(III)][HCrO_4^-]}{1 + K[As(III)]}$$

where k and K are defined by the equations

$$\mathrm{As(III) + HCrO_4^-} \underset{}{\overset{K}{\rightleftharpoons}} \mathrm{As(III).HCrO_4^-} \qquad \text{pre-equilibrium}$$

$$\mathrm{As(III).HCrO_4^-} \xrightarrow{k} \text{products} \qquad \text{rate-determining}$$

Under conditions of high Cr(VI) and low As(III) the results indicate that the dimer $Cr_2O_7^{2-}$ takes over a large control of the reaction. It seems clear that the reaction proceeds without the intervention of As(IV), and that As(V) and Cr(IV) are formed by oxygen transfer in the rate-controlling step (4). It is relevant that significant oxygen transfer occurs in the oxidation of U(IV) to U(VI) by Cr(VI).

Formation of peroxo complex of chromium(VI)

Chromium(VI) reacts with hydrogen peroxide in acid solution to form a blue peroxo complex of composition $CrO_5 . H_2O$. In the range of acid concentrations 0·03 to 6·0 M the rate of formation (measured by a flow method at 580 mμ) of the peroxo species is given by[29]

$$\text{rate} = k[HCrO_4^-][H_2O_2][H^+]$$

Depending on the source of the acid dependence, two mechanisms are possible. Firstly, the rate-determining stage may be identified as a reaction between H_2CrO_4 and peroxide, the former being in equilibrium with its conjugate acid $HCrO_4^-$

$$HCrO_4^- + H^+ \overset{K_1}{\rightleftharpoons} H_2CrO_4$$

$$H_2CrO_4 + H_2O_2 \xrightarrow{k_1} H_2CrO_5 + H_2O \quad \text{slow} \qquad (1)$$

$$H_2CrO_5 + H_2O_2 \rightarrow CrO_5 . H_2O + H_2O \quad \text{fast}$$

Alternatively, the acid dependence may originate from the reaction in the slow step of $H_3O_2^+$, formed by protonation of peroxide

$$H_2O_2 + H^+ \overset{K_2}{\rightleftharpoons} H_3O_2^+$$

$$H_3O_2^+ + HCrO_4^- \xrightarrow{k_2} H_2CrO_5 + H_2O \quad \text{slow} \qquad (2)$$

$$H_2CrO_5 + H_2O_2 \rightarrow CrO_5 . H_2O + H_2O \quad \text{fast}$$

Because the formation of H_2CrO_4 reaches a saturation value at high acidities, the rate of formation of $CrO_5 . H_2O$ deviates from a first-order dependence in hydrogen-ion concentration under these conditions and, at 4°, the rate law is then better expressed as

$$k_{obs} = \frac{5{\cdot}0 \times 10^3[H_2O_2][H^+]}{(1 + 0{\cdot}1[H^+])}$$

In the case of mechanism (1)

$$\frac{d[CrO_5]}{dt} = k_1[H_2CrO_4][H_2O_2]$$

$$= k_{obs}([HCrO_4^-] + [H_2CrO_4])$$

and
$$k' = \frac{k_{obs}}{[H_2O_2]} = \frac{k_1[H_2CrO_4]}{[HCrO_4^-] + [H_2CrO_4]}$$

therefore
$$\frac{1}{k'} = \frac{1}{k_1} + \frac{[HCrO_4^-]}{k_1[H_2CrO_4]}$$

or
$$\frac{1}{k'} = \frac{1}{k_1} + \frac{1}{k_1 K_1[H^+]}$$

For mechanism (2)

$$\frac{1}{k'} = \frac{K_1}{k_2 K_2} + \frac{1}{k_2 K_2[H^+]}$$

Thus for either mechanism to hold, plots of $1/k'$ versus $1/[H^+]$ should be linear, K_1 being obtainable from the intercept/slope value. The value of K_1 obtained in this fashion depends upon the medium and is $0{\cdot}35\ M^{-1}$ at 4° and $\mu = 6{\cdot}0$ M in perchloric acid, and $0{\cdot}1\ M^{-1}$ at 4° and $\mu = 6{\cdot}0$ M in nitric acid. The former value agrees well with K_1 obtained by direct spectrophotometry. If mechanism (1) were operative the intercept is to be identified with $1/k_1$, and k_1 is obtained as $2{\cdot}5 \times 10^4\ M^{-1}\ s^{-1}$ at 4°. If mechanism (2) were correct then k_2 is given by $K_1/(\text{intercept})K_2$ and, by using an estimated value of $2 \times 10^{-5}\ M^{-1}$ for K_2, then k_2 follows as $\sim 5 \times 10^8\ M^{-1}\ s^{-1}$. This last result is improbably high and suggests that the formation of the peroxo species occurs through mechanism (1).

The oxidation of iodide by permanganate

The oxidation of iodide by permanganate in acidic media has been the subject of a recent investigation.[30] The rate of the reaction

$$MnO_4^- + 5I^- + 8H^+ \rightarrow Mn(II) + \tfrac{5}{2}I_2 + 4H_2O$$

was studied at 35° by following the disappearance of permanganate spectrophotometrically at 520 mμ using a rapid-mixing device, the change in absorbance during the kinetic runs being measured by displaying the amplified signal from the photomultiplier on an oscilloscope and photographing the trace. Dependence of the rate on hydrogen-ion activity (a_{H^+}) over the pH range 3·2 to 6·2 is given by

$$\text{rate} = [MnO_4^-][I^-](k_2 + k_3 a_{H^+})$$

At constant pH the reaction is of simple second order with an observed rate constant defined by $k_2' = k_2 + k_3 a_{H^+}$. The mechanism

$$MnO_4^- + I^- \rightleftharpoons [O_3MnOI]^{2-} \qquad \text{rapid}$$

$$[O_3MnOI]^{2-} + HOH \rightarrow HOI + HMnO_4^{2-} \qquad \text{rate-determining}$$

$$HOI + H_3O^+ + 2I^- \rightarrow I_3^- + 2H_2O \qquad \text{rapid and complex}$$

is consistent with the kinetics of the hydrogen-ion independent path. The hydrogen-ion dependent path has a similar mechanism

$$MnO_4^- + I^- \rightleftharpoons [O_3MnOI]^{2-} \qquad \text{rapid}$$

$$[O_3MnOI]^{2-} + H_3O^+ \rightarrow HOI + H_2MnO_4^- \qquad \text{rate-determining}$$

It is interesting that the change from second-order to third-order behaviour as the pH is decreased can be viewed in terms of a competition for the intermediate between H_2O and H_3O^+ as electrophiles.

The permanganate–oxalate reaction

Elucidation of the mechanistic details of the permanganate+oxalate reaction

$$2MnO_4^- + 5C_2O_4^{2-} + 16H^+ \rightarrow 2Mn(II) + 10CO_2 + 8H_2O$$

has proved a difficult, although fascinating, task. It is clear that permanganate does not react directly with oxalate: early work showed that no reaction occurred between the two species for up to a week if traces of free Mn(II) were removed by the presence of an excess of fluoride ion. On the other hand, added Mn(II) has an accelerating effect on the reaction. Formation of a mono-oxalato complex of Mn(II) has been demonstrated by solubility and spectrophotometric measurements

$$Mn(II) + C_2O_4^{2-} \rightleftharpoons Mn(C_2O_4)$$

and the initiation of permanganate reduction is ascribed to[31]

$$Mn^{VII}O_4^- + Mn^{II}C_2O_4 \rightarrow Mn^{VI}O_4^{2-} + Mn^{III}C_2O_4^+$$

If Mn(II) ion is present in appreciable concentrations, manganate is reduced further by

$$Mn(VI) + Mn(II) \rightarrow 2Mn(IV)$$

$$Mn(IV) + Mn(II) \rightarrow 2Mn(III)$$

preferentially to disproportionation. In the absence of significant quantities of Mn(II), manganate is reduced, less favourably, by oxalate ion as indicated by

$$Mn(VI) + C_2O_4^{2-} \rightarrow Mn(IV) + 2CO_2$$

$$2Mn(IV) + C_2O_4^{2-} \rightarrow 2Mn(III) + 2CO_2$$

Mn(III) formed in either of these routes is destroyed by

$$Mn(C_2O_4)_n^{(3-2n)} \xrightarrow{\text{slow}} Mn(II) + (n-1)C_2O_4^{2-} + CO_2 + CO_2^-$$

$$Mn(C_2O_4)_n^{(3-2n)} + CO_2^- \rightarrow Mn(II) + nC_2O_4^{2-} + CO_2$$

where $n = 1$, 2, or 3. At extremely low concentrations of oxalate an autocatalytic step is introduced in the form of a reaction between permanganate and a Mn(III)–oxalate complex

$$MnO_4^- + MnC_2O_4^+ \rightarrow MnO_4^{2-} + Mn(IV)$$

The above discussion applies strictly to reaction in the absence of atmospheric oxygen: oxygen has been shown to cause the induced oxidation of oxalate, essentially by the intervention of a peroxo species (compare p.192)

$$CO_2^- + O_2 \rightleftharpoons O_2CO_2^-$$

which is then reduced to hydrogen peroxide

$$O_2CO_2^- + Mn(II) + 2H^+ \rightarrow Mn(III) + H_2O_2 + CO_2$$

The oxidation of cyanide by permanganate

Permanganate is reduced quantitatively by cyanide in strongly basic solution ($\geqslant$ pH 12) according to

$$2MnO_4^- + CN^- + 2OH^- \rightarrow 2MnO_4^{2-} + CNO^- + H_2O$$

As the pH is reduced below a value of 12 the reaction becomes increasingly non-stoichiometric and variable quantities of cyanogen and carbon dioxide are formed in addition to cyanate. Cyanogen, which can be isolated at pH 6–9, is presumed to occur in more alkaline solution but is rapidly hydrolysed under these conditions

$$2MnO_4^- + 6CN^- + 4H_2O \rightarrow 3(CN)_2 + 2MnO_2 + 8OH^-$$

$$(CN)_2 + 2OH^- \rightarrow CN^- + CNO^- + H_2O$$

Carbon dioxide is formed by the oxidation of cyanogen. The rate of oxidation of cyanide by permanganate is markedly affected by pH; in acid solution (pH < 6) the reaction is so slow as to be negligible. Above pH 7 the rate increases rapidly, achieving a maximum value at pH 9·1, and beyond this decreases again to a pH-independent value at pH $\geqslant 13$.

Two reaction paths are indicated.[32] Firstly, in strongly basic solutions, the rate expression

$$-\mathrm{d}[MnO_4^-]/\mathrm{d}t = k_a[MnO_4^-][CN^-]$$

applies, and tracer experiments with ^{18}O-labelled permanganate have shown that most of the oxygen introduced into the cyanate product originates in the permanganate and not in the solvent. The results receive rationalization in terms of a mechanism incurring oxygen transfer from permanganate to cyanide, thereby generating Mn(V) as an intermediate

$$MnO_4^- + CN^- \rightarrow [O_3Mn\text{---}O\text{---}CN]^{2-} \rightarrow MnO_3^- + OCN^-$$

followed by

$$MnO_3^- + MnO_4^- + 2OH^- \rightarrow 2MnO_4^{2-} + H_2O$$

Secondly, in less basic solutions, the rate law is more complicated

$$-\mathrm{d}[MnO_4^-]/\mathrm{d}t = k_b'[MnO_4^-][CN^-]^2[H^+]$$

$$= k_b[MnO_4^-][HCN][CN^-]$$

and the postulated path now involves the species $H(CN)_2^-$ formed by

$$HCN + CN^- \rightleftharpoons H(CN)_2^-$$

which can be oxidized by permanganate via a hydride transfer mechanism

$$MnO_4^- + H\text{—}(CN)_2^- \rightarrow HMnO_4^{2-} \text{ (or } H^+ + MnO_4^{3-}) + (CN)_2$$

This is followed by the formation of Mn(IV) (as MnO_2), through the decomposition of Mn(V), and also by hydrolysis of cyanogen to cyanate. The lack of appreciable amounts of oxygen-18 in the cyanate product supports this interpretation.

Further reactions of alkaline permanganate

Permanganate is unstable in strongly alkaline solution and decomposes to manganate and oxygen according to

$$4MnO_4^- + 4OH^- \rightarrow 4MnO_4^{2-} + 2H_2O + O_2$$

The reaction is best described[33] by a scheme comprising a number of simple one-electron and proton transfers and involving, as intermediates, OH, O^-, HO_2^-, HO_2, and O_2^-

$$MnO_4^- + OH^- \rightleftharpoons MnO_4^{2-} + OH$$

$$OH + OH^- \rightleftharpoons O^- + H_2O$$

$$MnO_4^- + O^- + OH^- \rightleftharpoons MnO_4^{2-} + HO_2^-$$

$$MnO_4^- + HO_2^- \rightleftharpoons MnO_4^{2-} + HO_2$$

$$HO_2 + OH^- \rightleftharpoons O_2^- + H_2O$$

$$MnO_4^- + O_2^- \rightarrow MnO_4^{2-} + O_2$$

The major criticism of the scheme is the doubt cast over the feasibility of the initial step. Energetically, the reaction is unlikely since the calculated standard free energy change is strongly positive ($\Delta G° \sim +33$ kcal). Furthermore, the apparent value given to the forward rate constant would necessitate an extremely rapid rate for the reverse reaction. The step may still be acceptable (with some reservations) if it is borne in mind that the hydroxide ion has an exceptionally high mobility in water. As a consequence it may be postulated that, in strongly alkaline solution, neighbouring water molecules can be transformed into hydroxide ions at will, by a Grotthus-chain transfer of protons. Support is lent to the above mechanism by the observation that, when the solvent is labelled $H_2{}^{18}O$, all the product oxygen is derived from the solvent.[34]

Further and convincing confirmation of the scheme comes from a kinetic study[35] of the oxidation of certain organic compounds RH (sodium isobutyrate and t-butanol) by alkaline solutions of permanganate. These compounds are not affected by neutral or dilute alkaline solutions of permanganate but are readily oxidized at room temperature by strongly alkaline permanganate. Furthermore, in the presence of a large excess of RH, oxygen evolution is suppressed. Formulating the first stages of the induced reaction as

$$MnO_4^- + OH^- \underset{k_2}{\overset{k_1}{\rightleftharpoons}} MnO_4^{2-} + OH$$

$$RH + OH \xrightarrow{k_3} R + H_2O$$

and applying the stationary-state approximation to the concentration of the hydroxy radical, produces the expression

$$\frac{-d[MnO_4^-]}{dt} = \frac{k_1 k_3[RH][MnO_4^-][OH^-]}{k_2[MnO_4^{2-}] + k_3[RH]}$$

This equation is adhered to closely over a range of concentrations; the value obtained for k_1, largely independent of the organic substrate, shows agreement with that derived from studies performed in the absence of substrate.

The isotopic exchange between permanganate and manganate in 0·16 M sodium hydroxide solution obeys the rate law[36]

$$\text{rate} = k[MnO_4^-][MnO_4^{2-}]$$

where $k = 710 \pm 30\ M^{-1}\ s^{-1}$ at 0°. Alkali metal ions are shown to have a marked accelerating effect on the rate of exchange, the order being given by $Cs^+ > K^+ > Na^+, Li^+$. As the reaction is clearly of the outer-sphere type, this result suggests that, in the absence of salt effects, the alkali metal ion is incorporated into the bridged activated complex, for example, $[MnO_4—Cs—MnO_4]^{2-}$. The kinetic results of the tracer and two independent nuclear magnetic resonance studies[37] are in reasonable agreement. Catalysis by $[Fe(CN)_6]^{3-}$ is ascribed to the establishment of the redox reaction

$$[Fe(CN)_6]^{3-} + MnO_4^{2-} \rightleftharpoons [Fe(CN)_6]^{4-} + MnO_4^-$$

The value estimated for the rate constant of the backward step is consistent with that determined by a flow method.

Reactions of manganate

The kinetics of the periodate+manganate reaction in alkaline solution are interesting since changes of two in the oxidation state of iodine, and of one in manganese, are involved

$$2MnO_4^{2-} + H_3IO_6^{2-} \rightarrow 2MnO_4^- + IO_3^- + 3OH^-$$

The kinetic data, obtained spectrophotometrically, indicate a second-order dependence on MnO_4^{2-} and are essentially in agreement with a scheme in which Mn(V) participates as an intermediate[38]

$$2MnO_4^{2-} \underset{k_2}{\overset{k_1}{\rightleftharpoons}} MnO_4^- + MnO_4^{3-}$$

$$MnO_4^{3-} + H_3IO_6^{2-} \xrightarrow{k_3} MnO_4^- + IO_3^- + 3OH^-$$

As the rate shows no retarding effect of MnO_4^- it follows that k_3 is considerably larger than k_2. Added MnO_4^- causes a reduction in rate with step k_2 competing against step k_3. From an analysis of the results, values for the individual rate constants were obtained: $k_1 = 107$, $k_2 = 2{\cdot}8 \times 10^6$, and $k_3 = 2 \times 10^7$ M^{-1} min^{-1}, at 35° and at an ionic strength of $\mu = 0{\cdot}1$ M. The surprisingly large value for k_3, the rate constant for the reaction between triply-charged hypomanganate and the doubly-charged iodate ion, would seem to indicate that MnO_4^{3-} hydrolyses to give ions of lower charge ($HMnO_4^{2-}$ and $H_2MnO_4^-$). The occurrence of a hydrogen-ion dependent term in the rate law suggests that manganate also is partially hydrolysed and that the first stage may be better expressed as

$$HMnO_4^- + MnO_4^{2-} \rightleftharpoons MnO_4^- + HMnO_4^{2-}$$

The non-complementary reaction between hypochlorite and manganate, represented overall by

$$2MnO_4^{2-} + OCl^- + H_2O \rightarrow 2MnO_4^- + Cl^- + 2OH^-$$

shows features to be expected for a one-equivalent–two-equivalent process. The rate law is complex[39]

$$\frac{-d[MnO_4^{2-}]}{dt} = \frac{k'[MnO_4^{2-}]^2[OCl^-]}{[MnO_4^-][OH^-]^2}$$

The inverse hydroxide-ion dependence is fairly conclusive evidence for a mechanism in which hydrolysed species are involved

$$MnO_4^{2-} + H_2O \rightleftharpoons HMnO_4^- + OH^-$$

$$2HMnO_4^- \rightleftharpoons H_2MnO_4^- + MnO_4^-$$

$$H_2MnO_4^- + OCl^- \rightarrow MnO_4^- + Cl^- + H_2O$$

References

1. M. Anbar and R. Rein, *J. Amer. Chem. Soc.*, 1959, **81**, 1813.
2. M. Anbar and H. Taube, *J. Amer. Chem. Soc.*, 1958, **80**, 1073.
3. M. W. Lister, *Canad. J. Chem.*, 1956, **34**, 465, 479.

4. M. W. Lister and P. Rosenblum, *Canad. J. Chem.*, 1961, **39**, 1645.
5. M. W. Lister and P. Rosenblum, *Canad. J. Chem.*, 1963, **41**, 3013; Y. T. Chia and R. E. Connick, *J. Phys. Chem.*, 1959, **63**, 1518.
6. T. C. Hoering, R. C. Butler, and H. O. McDonald, *J. Amer. Chem. Soc.*, 1956, **78**, 4829; M. Anbar and S. Guttmann, *J. Amer. Chem. Soc.*, 1961, **83**, 4741.
7. T. C. Hoering, F. T. Ishimori, and H. O. McDonald, *J. Amer. Chem. Soc.*, 1958, **80**, 3876.
8. M. Anbar and S. Guttmann, *J. Amer. Chem. Soc.*, 1961, **83**, 781.
9. I. Pecht and Z. Luz, *J. Amer. Chem. Soc.*, 1965, **87**, 4068.
10. K. Kustin and E. C. Lieberman, *J. Phys. Chem.*, 1964, **68**, 3869.
11. Z. Luz and I. Pecht, *J. Amer. Chem. Soc.*, 1966, **88**, 1152.
12. J. Reuben, D. Samuel, H. Selig, and J. Shamir, *Proc. Chem. Soc.*, 1963, 270.
13. D. M. Kern and C. H. Kim, *J. Amer. Chem. Soc.*, 1965, **87**, 5309.
14. K. J. Morgan, M. G. Peard, and C. F. Cullis, *J. Chem. Soc.*, 1951, 1865; K. J. Morgan, *Quart. Rev.*, 1954, **8**, 123.
15. D. E. Myers and J. W. Kennedy, *J. Amer. Chem. Soc.*, 1950, **72**, 897.
16. A. Indelli, F. Ferranti, and F. Secco, *J. Phys. Chem.*, 1966, **70**, 631.
17. J. Halperin and H. Taube, *J. Amer. Chem. Soc.*, 1950, **72**, 3319; 1952, **74**, 375.
18. E. R. Nightingale, *J. Phys. Chem.*, 1960, **64**, 162.
19. J. Halperin and H. Taube, *J. Amer. Chem. Soc.*, 1952, **74**, 380.
20. See, for example, I. Johnson and V. K. La Mer, *J. Amer. Chem. Soc.*, 1947, **69**, 1184; E. M. Zaiser and V. K. La Mer, *J. Colloid Sci.*, 1948, **3**, 571.
21. R. E. Davis, *J. Amer. Chem. Soc.*, 1958, **80**, 3565.
22. R. E. Davis, *J. Phys. Chem.*, 1958, **62**, 1599.
23. G. P. Haight, E. Perchonock, F. Emmenegger, and G. Gordon, *J. Amer. Chem. Soc.*, 1965, **87**, 3835.
24. C. Altman and E. L. King, *J. Amer. Chem. Soc.*, 1961, **83**, 2825.
25. E. L. King and J. A. Neptune, *J. Amer. Chem. Soc.*, 1955, **77**, 3186.
26. F. H. Westheimer, *Chem. Rev.*, 1949, **45**, 419.
27. J. H. Espenson and E. L. King, *J. Amer. Chem. Soc.*, 1963, **85**, 3328.
28. J. G. Mason and A. D. Kowalak, *Inorg. Chem.*, 1964, **3**, 1248.
29. M. Orhanović and R. G. Wilkins, *J. Amer. Chem. Soc.*, 1967, **89**, 278.
30. L. J. Kirschenbaum and J. R. Sutter, *J. Phys. Chem.*, 1966, **70**, 3863.
31. S. J. Adler and R. M. Noyes, *J. Amer. Chem. Soc.*, 1955, **77**, 2036.
32. R. Stewart and R. Van der Linden, *Canad. J. Chem.*, 1960, **38**, 2237.
33. M. C. R. Symons, *J. Chem. Soc.*, 1953, 3956.
34. M. C. R. Symons, *J. Chem. Soc.*, 1954, 3676.
35. K. A. K. Lott and M. C. R. Symons, *Disc. Faraday Soc.*, 1960, **29**, 205.
36. J. C. Sheppard and A. C. Wahl, *J. Amer. Chem. Soc.*, 1957, **79**, 1020; L. Gjertsen and A. C. Wahl, *J. Amer. Chem. Soc.*, 1959, **81**, 1572.
37. A. D. Britt and W. M. Yen, *J. Amer. Chem. Soc.*, 1961, **83**, 4516; O. E. Myers and J. C. Sheppard, *J. Amer. Chem. Soc.*, 1961, **83**, 4739.
38. M. W. Lister and Y. Yoshino, *Canad. J. Chem.*, 1960, **38**, 2342.
39. M. W. Lister and Y. Yoshino, *Canad. J. Chem.*, 1961, **39**, 96.

Bibliography

J. O. Edwards, Rate Laws and Mechanisms of Oxyanion Reactions with Bases, in *Chemical Reviews*, 1952, **50**, 455.

W. A. Waters, Mechanisms of Oxidation by Compounds of Chromium and Manganese, in *Quarterly Reviews*, 1958, **12**, 277.

F. H. Westheimer, The Mechanisms of Chromic Acid Oxidations, in *Chemical Reviews*, 1949, **45**, 419.

J. W. Ladbury and C. F. Cullis, Kinetics and Mechanism of Oxidation by Permanganate, in *Chemical Reviews*, 1958, **58**, 403.

A. Carrington and M. C. R. Symons, Structure and Reactivity of the Oxyanions of Transition Metals, in *Chemical Reviews*, 1963, **63**, 443.

R. Stewart, Oxidation by Permanganate, in *Oxidation in Organic Chemistry* (ed. K. B. Wiberg), p. 1, Academic Press, 1965.

5. Free radical reactions

This chapter is concerned with some authenticated examples of free radical reactions in solution involving the peroxodisulphate (persulphate) ion and hydrogen peroxide. Both are strong oxidizing agents, the standard electrode potential of $S_2O_8^{2-}(aq) + 2e^- \rightarrow 2SO_4^{2-}(aq)$ being given as $E° = +2{\cdot}0$ V; and both possess O—O bonds, peroxodisulphate having a structure $^-OSO_2O—OSO_2O^-$, analogous to that of the diacyl peroxides. The treatment covers, firstly, the thermal decomposition of peroxodisulphate. This is followed by some second-order reactions of peroxodisulphate with iron(II) and some chain processes with oxalate, thiosulphate, and alcohol. The peroxide–peroxodisulphate chain mechanism is then discussed and finally the catalytic decomposition of hydrogen peroxide.

The thermal decomposition of peroxodisulphate

At a given pH, in solutions buffered against accumulation of hydrogen ions, the thermal decomposition of peroxodisulphate conforms accurately to a first-order rate law[1]

$$-d[S_2O_8^{2-}]/dt = k_{obs}[S_2O_8^{2-}]$$

The observed rate constant, k_{obs}, is pH-dependent[2] and has the form $k_{obs} = k_0 + k_{H^+}[H^+]$. In the range of pH 3–13, only the k_0 path contributes to the rate, and under these conditions the decomposition corresponds to the stoichiometry

$$2S_2O_8^{2-} + 2H_2O \rightarrow 4HSO_4^- + O_2$$

The generally accepted mechanism for the hydrogen-ion independent path involves, as the primary step, the fission of the O—O bond to give sulphate-ion radicals

$$S_2O_8^{2-} \xrightarrow{k_1} 2SO_4^- \qquad \text{rate-determining} \qquad (1)$$

The observed overall activation energy for the peroxodisulphate decomposition of 33·5 kcal agrees well with the O—O bond strength.

The SO_4^- radicals then attack the solvent to generate hydroxyl radicals

$$SO_4^- + H_2O \xrightarrow{k_2} HSO_4^- + OH \tag{2}$$

followed by the destruction of OH radicals as represented by the stoichiometric equation

$$2OH \rightarrow H_2O + \tfrac{1}{2}O_2 \tag{3}$$

This oxygen-formation reaction may well involve hydrogen peroxide as an intermediate

$$2OH \rightarrow H_2O_2 \tag{4}$$

which reacts either as

$$H_2O_2 \rightarrow H_2O + \tfrac{1}{2}O_2 \tag{5}$$

or by a chain process (see p. 199) represented by

$$H_2O_2 + S_2O_8^{2-} \rightarrow 2HSO_4^- + O_2 \tag{6}$$

Steps (2) to (6) are compatible with the evidence of oxygen-18 studies which show that the oxygen produced in the reaction in alkaline solution originates entirely from the solvent. However, it has been pointed out that a chain mechanism is also in accord with the observed kinetics.[1] This is made up of steps (1) and (2) together with

$$OH + S_2O_8^{2-} \xrightarrow{k_4} HSO_4^- + SO_4^- + \tfrac{1}{2}O_2$$

$$SO_4^- + OH \xrightarrow{k_5} HSO_4^- + \tfrac{1}{2}O_2$$

Application of the steady-state approximation to the concentrations of OH and SO_4^- radicals leads to a rate law showing first-order dependence on $S_2O_8^{2-}$ (see p. 10; but note that the step k_3 given there is absent for water as substrate).

An alternative rate-determining step to (1) has been suggested,[3] assuming the formation of only one sulphate radical

$$S_2O_8^{2-} + H_2O \rightarrow HSO_4^- + SO_4^- + OH \tag{7}$$

However, reaction (7) is not in accord with the results obtained in the peroxodisulphate-initiated polymerization of styrene monomer.[4] In this system the growing polymer molecule captures sulphate radicals in competition with the solvent. A comparison has been made of the rate of capture of sulphate radicals by the polymer with the rate of production of such radicals assuming reaction (1) *or* reaction (7). In one experiment the rate of incorporation of sulphate radicals in the polymer was assessed as 6–7 × 10^{12} radicals cm^{-3} s^{-1} and the rate of generation as 8·8 × 10^{12} by reaction (1). At least in these circumstances reaction (7) cannot be responsible for the generation of sulphate radicals since it would have a rate of 4·4 × 10^{12} radicals cm^{-3} s^{-1} (i.e., half that for reaction (1)), a value which is less than the rate of sulphate-radical capture. The proposal that the hydrogen-ion independent path is initiated by the rapid equilibriation

$$S_2O_8^{2-} \rightleftharpoons SO_4^{2-} + SO_4$$

is untenable also, as exchange of sulphur-35 does not take place between sulphate ion and peroxodisulphate. This observation demonstrates also that reaction (1) itself is irreversible.

The hydrogen-ion dependent (k_{H^+}) path predominates at a pH less than 3 and the suggestion has been made[2] that it proceeds by

$$S_2O_8^{2-} + H^+ \rightarrow HSO_4^- + SO_4 \quad \text{rate-determining} \quad (8)$$

$$SO_4 \rightarrow SO_3 + \tfrac{1}{2}O_2 \quad (9)$$

$$SO_3 + H_2O \rightarrow H_2SO_4 \quad (10)$$

$$SO_4 + H_2O \rightarrow H_2SO_5 \quad (11)$$

It should be noted that reaction (9) has a stoichiometric significance only. Reaction (11) assumes importance as the acidity increases and, at a hydrogen-ion concentration of greater than 2 M, peroxomonosulphuric acid is formed exclusively. The mechanism offers an explanation of the results of oxygen-18 studies which have shown that all oxygen produced in the reaction under acid conditions comes from the peroxodisulphate.

The Fe(II)–peroxodisulphate reaction

The reaction of iron(II) with peroxodisulphate has been formulated[5] in terms of a two-stage process, the first and rate-determining step being the formation of a sulphate free radical

$$Fe(II) + S_2O_8^{2-} \rightarrow Fe(III) + SO_4^- + SO_4^{2-} \quad (1)$$

The sulphate radical then oxidizes another ferrous ion

$$Fe(II) + SO_4^- \xrightarrow{k_2} Fe(III) + SO_4^{2-} \quad (2)$$

The sulphate free radicals are able to initiate oxidations and polymerizations of organic compounds, and their presence is confirmed by the incorporation of sulphur-35, from ^{35}S-labelled peroxodisulphate, into polymer chains. Also studies[6] of the competition of ferrous ion and methyl acrylate for the sulphate radical by reactions (2) and (3)

$$M + SO_4^- \xrightarrow{k_3} MSO_4^- \qquad (3)$$

lead to a value for k_3/k_2 of 3×10^{-3} at 25° whereas the competition ratio for the hydroxyl radical, generated in the Fe(II)+hydrogen peroxide system, has a value of 5·0 under the same conditions. In the case of peroxodisulphate there seems little doubt, therefore, that the active intermediate is the sulphate radical and not the hydroxyl radical, produced by reaction of the former with water.

A stoichiometry of two ferrous ions oxidized to one ion of peroxodisulphate consumed results by combining reactions (1) and (2). Although this is realized when excess Fe(II) is added rapidly to peroxodisulphate, stoichiometries as low as 1·1 are observed at low concentrations of Fe(II). This result may be explained on the basis that in this case sulphate radicals are removed by reaction with the solvent, rather than by reaction (2).

The induced oxidation of As(III) by Fe(II)–peroxodisulphate

The peroxodisulphate–iron(II) couple has been shown to cause the induced oxidation of arsenic(III).[7] The reaction between peroxodisulphate and arsenic(III) in the absence of iron(II) is very slow, the induced oxidation being caused by the sulphate-ion radicals produced by

$$Fe(II) + S_2O_8^{2-} \xrightarrow[\text{slow}]{k} Fe(III) + SO_4^{2-} + SO_4^-$$

which then attack arsenic(III) to yield arsenic(IV) as an intermediate

$$As(III) + SO_4^- \xrightarrow{\text{fast}} As(IV) + SO_4^{2-}$$

Arsenic(IV) is then oxidized further by iron(III) thus regenerating iron(II)

$$As(IV) + Fe(III) \xrightarrow{\text{fast}} As(V) + Fe(II)$$

The stoichiometry of the overall reaction is

$$As(III) + S_2O_8^{2-} \rightarrow As(V) + 2SO_4^{2-}$$

with an induction factor

$$\left(=\frac{\text{equiv. As(III) oxidized}}{\text{equiv. Fe(II) oxidized}}\right)$$

of infinity in the presence of an excess of iron(III). The experimental rate law is

$$-\mathrm{d}[S_2O_8^{2-}]/\mathrm{d}t = k[S_2O_8^{2-}][Fe(II)]$$

with the rate constant, k, having the same value as that for the system $Fe(II) + S_2O_8^{2-}$ in the absence of As(III). As the ratio of arsenic(III) to iron(II) is reduced, a reduction in the value of the induction factor occurs and iron(II) competes with arsenic(III) for the sulphate free radical

$$Fe(II) + SO_4^- \rightarrow Fe(III) + SO_4^{2-}$$

When iron(III) is complexed with fluoride the induction factor approaches a limiting value of zero. In the absence of iron(III), therefore, the arsenic(IV) intermediate is postulated to react with iron(II) to reform arsenic(III)

$$As(IV) + Fe(II) \rightarrow As(III) + Fe(III)$$

The influence of copper(II) on the $Fe(II)–S_2O_8^{2-}$ induced oxidation of As(III) is similar to that of Fe(III) with an induction factor approaching a limiting value of infinity, explainable by the mechanism

$$As(IV) + Cu(II) \rightarrow As(V) + Cu(I)$$

with the fate of the intermediate copper(I), either

$$Cu(I) + Fe(III) \rightarrow Cu(II) + Fe(II)$$

or

$$Cu(I) + S_2O_8^{2-} \rightarrow Cu(II) + SO_4^{2-} + SO_4^-$$

The presence of dissolved oxygen has a drastic effect on the induced oxidation of arsenic(III) which now shows all the characteristics of a chain mechanism.[8] In the absence of oxygen, the sum $\frac{1}{2}R_{Fe} + R_{As} = 1$, where R_{Fe} is the ratio of moles Fe(II) oxidized to moles $S_2O_8^{2-}$ reduced and R_{As}, the ratio of moles As(III) oxidized to moles $S_2O_8^{2-}$ reduced. However, in the presence of oxygen, $\frac{1}{2}R_{Fe} + R_{As}$ can achieve a value of as high as 18·5, indicating a considerable chain oxidation. The influence of oxygen is interpretable in terms of the reaction of arsenic(IV) with dissolved oxygen to produce a peroxo radical

$$As(IV) + O_2 + H^+ \rightarrow As(V) + HO_2$$

which is capable of causing further oxidations by

$$Fe(II) + HO_2 + H^+ \rightarrow Fe(III) + H_2O_2$$

$$Fe(II) + H_2O_2 \rightarrow Fe(III) + OH + OH^-$$

$$As(III) + OH \rightarrow As(IV) + OH^-$$

thus regenerating the arsenic(IV) intermediate which can then produce further peroxo radicals. Chain termination takes place by

$$Fe(III) + HO_2 \rightarrow Fe(II) + O_2 + H^+$$

The oxidation of oxalate by peroxodisulphate

The oxidation of oxalate ion by peroxodisulphate is catalysed by copper(II) but the catalysed reaction is inhibited by the presence of oxygen.[9] The reaction has been investigated in the absence of oxygen and the kinetics follow a rate law which is first-order in peroxodisulphate, zero-order in oxalate, and half-order in copper(II). The reaction is greatly inhibited by allyl acetate and in the presence of this radical scavenger the rate becomes identical with the thermal decomposition of peroxodisulphate. The chain-initiating step is thus the homolytic fission of $S_2O_8^{2-}$. Inhibition by oxygen suggests that a chain reaction occurs in which radicals derived from the oxalate substrate take part. A chain mechanism of considerable length is indicated also by the general difficulties encountered in obtaining reproducible kinetic data, and the susceptibility of the system to the nature of the surface of the reaction vessel. One mechanism proposed is

Initiation:

$$S_2O_8^{2-} \xrightarrow{k_1} 2SO_4^- \qquad (1)$$

Propagation:

$$SO_4^- + Cu^{II}(C_2O_4)_2^{2-} \xrightarrow{k_2} SO_4^{2-} + Cu^{III}(C_2O_4)_2^- \qquad (2)$$

$$Cu^{III}(C_2O_4)^- \xrightarrow{k_3} Cu^{II}C_2O_4 + CO_2 + CO_2^- \qquad (3)$$

$$Cu^{II}C_2O_4 + C_2O_4^{2-} \xrightarrow{k_4} Cu^{II}(C_2O_4)^{2-} \qquad (4)$$

$$CO_2^- + S_2O_8^{2-} \xrightarrow{k_5} CO_2 + SO_4^{2-} + SO_4^- \qquad (5)$$

Termination:

$$CO_2^- + SO_4^- \xrightarrow{k_6} CO_2 + SO_4^{2-} \qquad (6)$$

with an overall stoichiometry of

$$C_2O_4^{2-} + S_2O_8^{2-} \rightarrow 2CO_2 + 2SO_4^{2-}$$

Making use of the steady-state approximation for the concentrations of radical-ions and copper(III) intermediate, the rate law can be derived

$$\frac{-d[S_2O_8^{2-}]}{dt} = \left(\frac{k_1 k_2 k_5}{k_6}\right)^{1/2} [S_2O_8^{2-}][Cu^{II}(C_2O_4)_2^{2-}]^{1/2}$$

assuming the inequality $4k_2 k_5[Cu^{II}(C_2O_4)_2^{2-}] \gg k_1 k_6$. This assumption is justified since under the conditions of the investigation k_1 is $2{\cdot}0 \times 10^{-7}$ s^{-1} at 40·5° and k_{obs}, defined as

$$\left(\frac{k_1 k_2 k_5}{k_6}\right)^{1/2} [Cu^{II}(C_2O_4^{2-}]^{1/2}$$

is 50 to 2000 times larger than k_1 at the same temperature, and therefore

$$(2k_{obs}/k_1)^2 = 4k_2 k_5[Cu^{II}(C_2O_4)_2^{2-}]/k_1 k_6 \gg 10^4$$

An alternative mechanism involving the hydroxyl radical has been postulated which includes the following steps in addition to reactions (1), (3), (4), and (5):

$$SO_4^- + H_2O \xrightarrow{k_7} SO_4^{2-} + H^+ + OH$$

$$OH + Cu^{II}(C_2O_4)_2^{2-} \xrightarrow{k_8} OH^- + Cu^{III}(C_2O_4)_2^-$$

$$H^+ + OH^- \xrightarrow{k_9} H_2O$$

$$OH + CO_2^- \xrightarrow{k_{10}} OH^- + CO_2$$

The derived rate law is identical with the previous one except that k_8 and k_{10} replace k_2 and k_6, respectively. Inhibition by oxygen is attributed to its reaction with CO_2^- radicals to form a peroxide complex

$$O_2 + CO_2^- \rightleftharpoons O_2CO_2^-$$

followed by termination of the chain by

$$O_2CO_2^- + SO_4^- \rightarrow O_2 + CO_2 + SO_4^{2-}$$

The first step is of necessity reversible since the reaction preserves the same form of rate law in the presence of oxygen.

The Ag(I)-catalysed oxidation of oxalate

The silver(I)-catalysed oxidation of oxalate ion by peroxodisulphate has been studied [10] in the absence of oxygen as a sequel to the Cu(II) catalysis, both reactions having the same stoichiometry but differing markedly in their kinetics. The empirical rate law for the silver(I) catalysis assumes one of two forms, depending on the concentration of substrate:

(*a*)

$$\text{for } [S_2O_8^{2-}] > 0{\cdot}004 \text{ M}, \; -\mathrm{d}[S_2O_8^{2-}]/\mathrm{d}t = k_a[S_2O_8^{2-}]^{3/2}[Ag(I)]^{1/2}$$

(*b*)

$$\text{for } [S_2O_8^{2-}] \leqslant 0{\cdot}004 \text{ M}, \; -\mathrm{d}[S_2O_8^{2-}]/\mathrm{d}t = k_b[S_2O_8^{2-}]^2$$

The presence of a chain mechanism is indicated by these results and by the inhibitory effect of allyl acetate and oxygen. In addition the reaction is very much faster than other silver(I)-catalysed processes of peroxodisulphate which are known to be of the non-chain type.

The following mechanism accounts for both forms of rate law:

Initiation:

$$\mathrm{Ag(I)} + \mathrm{S_2O_8^{2-}} \xrightarrow{k_1} \mathrm{Ag(II)} + \mathrm{SO_4^-} + \mathrm{SO_4^{2-}} \qquad (1)$$

Propagation:

$$\mathrm{SO_4^-} + \mathrm{Ag(I)} \xrightarrow{k_2} \mathrm{SO_4^{2-}} + \mathrm{Ag(II)} \qquad (2)$$

$$\mathrm{Ag(II)} + \mathrm{C_2O_4^{2-}} \xrightarrow{k_3} \mathrm{Ag(I)} + \mathrm{CO_2} + \mathrm{CO_2^-} \qquad (3)$$

$$\mathrm{CO_2^-} + \mathrm{S_2O_8^{2-}} \xrightarrow{k_4} \mathrm{CO_2} + \mathrm{SO_4^-} + \mathrm{SO_4^{2-}} \qquad (4)$$

Termination:

$$2\mathrm{CO_2^-} \xrightarrow{k_5} \mathrm{C_2O_4^{2-}} \qquad (5)$$

$$\mathrm{CO_2^-} + \mathrm{Ag(I)} \xrightarrow{k_6} \mathrm{CO_2} + \mathrm{Ag} \qquad (6)$$

The nature of Ag(I) depends upon the pH of the oxalate solution with complexing, as $[Ag^{I}C_2O_4]^-$, complete at high pH. From the usual steady-state approximation and the assumption that

$$k_1[Ag(I)][S_2O_8^{2-}] \ll k_4[CO_2^-][S_2O_8^{2-}]$$

(as required for a long chain), it can be shown that

$$\frac{-d[S_2O_8^{2-}]}{dt} = \frac{k_4 k_6}{2k_5}[Ag(I)][S_2O_8^{2-}]\{(1+X)^{1/2}-1\}$$

where

$$X = \frac{8k_1 k_5[S_2O_8^{2-}]}{k_6^2[Ag(I)]}$$

For concentrations of peroxodisulphate such that $X \gg 1$, the rate expression reduces to

$$\frac{-d[S_2O_8^{2-}]}{dt} = \left(\frac{2k_1 k_4^2}{k_5}\right)^{1/2}[S_2O_8^{2-}]^{3/2}[Ag(I)]^{1/2}$$

in agreement with rate law (*a*). For low peroxodisulphate concentrations where $X \ll 1$, rate law (*b*) is obtained

$$\frac{-d[S_2O_8^{2-}]}{dt} = \frac{2k_1 k_4}{k_6}[S_2O_8^{2-}]^2$$

by approximating $(1+X)^{1/2}$ as $1+\frac{1}{2}X$. Also the relative importance of the termination steps, (5) and (6), can be assessed since

$$\frac{k_5[CO_2^-]^2}{k_6[CO_2^-][Ag(I)]} = (1+X)^{1/2}-1$$

Under the conditions generating rate law (*a*), reaction (5) predominates, whereas for law (*b*), reaction (6) is the important one. The observed formation of metallic silver at the lower peroxodisulphate concentrations lends support to the argument.

The oxidation of thiosulphate by peroxodisulphate

The stoichiometry of this reaction corresponds essentially to

$$S_2O_8^{2-} + 2S_2O_3^{2-} \rightarrow S_4O_6^{2-} + 2SO_4^{2-}$$

although an unknown side-reaction does occur. The reaction is first-order in peroxodisulphate but is zero-order in thiosulphate concentration.[11] The rate is unaffected by oxygen, and by changes in hydrogen-ion concentration over the pH range 4 to 10·5. A radical chain mechanism is favoured for several reasons: for example, (*a*) the system is an efficient initiator of vinyl polymerizations, (*b*) well-defined periods of induction occur, and (*c*) under the same set of conditions the rate constant is very much larger than that for the thermal decomposition of peroxodisulphate, as well as being independent of thiosulphate concentration. The chain mechanism which has been proposed is

$$S_2O_8^{2-} \xrightarrow{k_1} 2SO_4^-$$

$$SO_4^- + H_2O \xrightarrow{k_2} HSO_4^- + OH$$

$$OH + S_2O_3^{2-} \xrightarrow{k_3} OH^- + S_2O_3^-$$

$$S_2O_3^- + S_2O_8^{2-} \xrightarrow{k_4} SO_4.S_2O_3^{2-} + SO_4^-$$

$$S_2O_3^- + SO_4^- \xrightarrow{k_5} SO_4.S_2O_3^{2-}$$

$$SO_4.S_2O_3^{2-} + S_2O_3^{2-} \xrightarrow{\text{fast}} S_4O_6^{2-} + SO_4^{2-}$$

From the usual steady-state approximation that the concentrations of the free radicals (SO_4^-, OH, and $S_2O_3^-$) are constant (see p. 11) and the assumption that k_1 is small, the rate law obtained is

$$\frac{-d[S_2O_8^{2-}]}{dt} = \frac{-d[S_2O_3^{2-}]}{dt} = \left(\frac{k_1 k_2 k_4}{k_5}\right)^{1/2} [S_2O_8^{2-}]$$

in agreement with the empirical rate expression. Other chain mechanisms consistent with the kinetics have the drawback of requiring a termolecular chain-termination step. One remarkable feature of the reaction is its extreme sensitivity to trace quantities of copper, and to a lesser extent, of iron. It appears very likely that a considerable factor in the reaction is the catalysis by minute amounts of these metal ions, present as impurities in the solvent and reagents. A mechanism for the copper-catalysed reaction is not available but it has been tentatively suggested that complexes of copper(I) and thiosulphate are responsible for the catalysis, or that copper(III) may be involved. It is somewhat unexpected that silver(I) is ineffective as a catalyst.

The oxidation of methanol by peroxodisulphate

The thermal decomposition of peroxodisulphate is accelerated in the presence of methanol; for example, the rate is increased by a factor of about 25 at 1 M concentration and 80° in a phosphate buffer of pH 8.[1] A change in order occurs in the presence of the alcohol and the reaction conforms now to a 3/2 order, as shown by the linearity of plots of $1/[S_2O_8^{2-}]^{1/2}$ versus time (Fig. 5.1). Furthermore, the rate constants obtained show an approximate dependence on the square

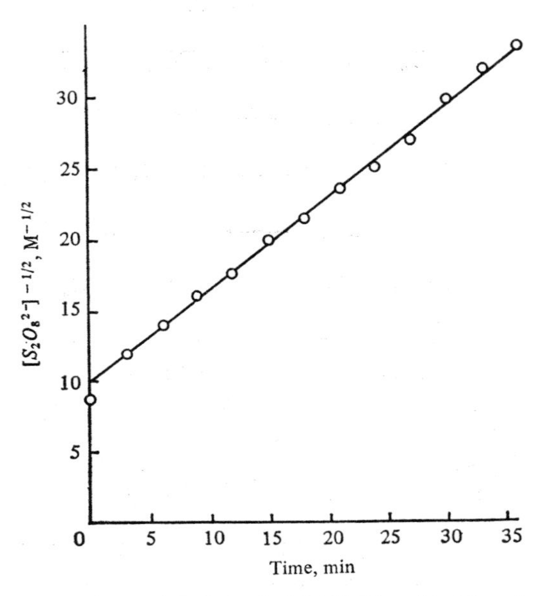

Fig. 5.1. Decomposition of $S_2O_8^{2-}$ (0·01375 M) at 79·8° and pH 8 in the presence of 0·488 M methanol, plotted as a 3/2 order reaction. From P. D. Bartlett and J. D. Cotman, *J. Amer. Chem. Soc.*, 1949, **71**, 1419.

root of the methanol concentration. A possible reaction scheme is

$$S_2O_8^{2-} \xrightarrow{k_{1a}} 2SO_4^- \tag{1a}$$

$$S_2O_8^{2-} + CH_3OH \xrightarrow{k_{1b}} HSO_4^- + SO_4^- + CH_2OH \tag{1b}$$

$$SO_4^- + CH_3OH \xrightarrow{k_2} HSO_4^- + CH_2OH \tag{2}$$

$$CH_2OH + S_2O_8^{2-} \xrightarrow{k_3} HSO_4^- + SO_4^- + HCHO \tag{3}$$

$$2CH_2OH \xrightarrow{k_4} CH_3OH + HCHO \tag{4}$$

Early work could not decide whether the primary step is reaction (1a) or (1b) but a detailed study[12] of the induced polymerization of allyl acetate has allowed a discrimination to be made. The kinetic results of this reaction display two main features. Firstly, the initial rate of decomposition of peroxodisulphate in the presence of allyl acetate is identical with the rate of thermal decomposition, and hence it follows that all radicals which initiate the polymerization of allyl acetate must derive from reaction (1a). Secondly, alcohols (e.g., methanol) accelerate the reaction, but in the additional presence of allyl acetate the rapid rate of disappearance of peroxodisulphate is reduced to that of the normal thermal decomposition. It is apparent therefore that alcohols do not react directly with peroxodisulphate by reaction (1b), and, as a result of the capture of nearly all sulphate-ion radicals by the monomer, the chain decomposition (propagated in reaction (3) by hydroxymethylene radicals derived from the alcohol) is practically eliminated. The sequence of reactions, represented by eqs. (1a), (2), (3), and (4), generate a complex rate law of the form

$$\frac{-\mathrm{d}[S_2O_8{}^{2-}]}{\mathrm{d}t} = k_{1a}[S_2O_8{}^{2-}] + k_3\left(\frac{k_{1a}}{k_4}\right)^{1/2}[S_2O_8{}^{2-}]^{3/2}$$

and the observed dependence of the rate on the square root of the methanol concentration has, until recently, remained unexplained. By testing the effect of the aldehyde product on the rate it has been found that the apparent kinetic order in alcohol is spurious and is due to a competition between alcohol and aldehyde for the oxidizing radicals, SO_4^- and/or OH.[13] In the absence of oxygen and with a constant amount of aldehyde present, the oxidations of methanol and ethanol by peroxodisulphate are both 3/2 order in $S_2O_8^{2-}$, and zero order in alcohol.

The kinetics of dissociation of peroxodisulphate have been investigated in ethanol–water mixtures by the use of the coloured free radical diphenyl picryl hydrazyl (DPPH) as a radical scavenger.[14] The rate of consumption of DPPH is independent of its concentration but is proportional to the concentration of peroxodisulphate. The decomposition was shown to be consistent with the scheme

$$S_2O_8^{2-} \xrightarrow{k_{1a}} 2SO_4^-$$

$$\text{OH or } SO_4^- + \text{DPPH} \rightarrow \text{stable products}$$

where $k_{1a} = 3{\cdot}7 \times 10^{13} \exp(-28390/RT)\ \mathrm{s}^{-1}$ in a 1:1 v/v alcohol–water solvent. No reaction with alcohol occurred since all SO_4^- radicals are rapidly removed by DPPH. The dissociation of peroxodisulphate is very susceptible to small amounts of silver ion, the catalytic path being

$$Ag(I) + S_2O_8^{2-} \xrightarrow{k_{1c}} Ag(II) + SO_4^{2-} + SO_4^{-}$$

$$Ag(II) + OH^{-} \rightarrow Ag(I) + OH$$

with $k_{1c} = 3{\cdot}1 \times 10^{11} \exp(-17900/RT)$ M^{-1} s^{-1}.

The hydrogen peroxide–peroxodisulphate reaction

The reaction between peroxodisulphate and hydrogen peroxide is the subject of an excellent paper by Tsao and Wilmarth.[15] Indeed, this investigation, with its concern over purity and commendable single-mindedness, sets a standard for further work on peroxide systems. To obtain results reproducible to within 10 per cent extreme precautions were necessary to remove trace impurities from the reagents and the solvent. The impurity level could be gauged from the length of the induction period, which in several experiments persisted as long as three to six hours. The water solvent was purified by redistillations over alkaline and acidic peroxodisulphate and all solid reagents were recrystallized at least twice from purified water. Inhibitor-free peroxide and triply-distilled perchloric acid were employed. The kinetic experiments were performed at 30° with peroxodisulphate concentrations determined iodometrically at pH 5·5–7·0 after peroxide had been removed with alkaline bromine water; peroxide was determined ceratimetrically after removal of peroxodisulphate by passage through a column of activated alumina.

The kinetic data conformed to the following complicated rate law

$$\frac{-\mathrm{d}[S_2O_8^{2-}]}{\mathrm{d}t} = \frac{[S_2O_8^{2-}]}{\left[\dfrac{9{\cdot}5 \times 10^6}{[H_2O_2]} + \dfrac{7{\cdot}6 \times 10^7[S_2O_8^{2-}]}{[H_2O_2]} + 7{\cdot}9 \times 10^8 + 2{\cdot}9 \times 10^{10}[S_2O_8^{2-}]\right]^{1/2}}$$

Fortunately, the relative values of the numerical constants in the denominator of this expression differ considerably, and under the appropriate experimental conditions the first, third, or fourth term assumes predominance, resulting in an overall order of three-halves (*a*), one (*b*), and one-half (*c*), respectively:

(*a*) $\text{rate} = k_a[S_2O_8^{2-}][H_2O_2]^{1/2}$

(*b*) $\text{rate} = k_b[S_2O_8^{2-}]$

and (*c*) $\text{rate} = k_c[S_2O_8^{2-}]^{1/2}$

Initially the rate shows a linear dependence on $[H_2O_2]^{1/2}$ given by rate law (*a*), but then achieves a limiting value at peroxide concentrations greater than 0·025 M. Above this limit rate laws (*b*) and (*c*) apply, the former at low peroxodisulphate concentrations and the latter at higher concentrations. The reaction has been discussed in terms of a chain mechanism, with homolytic fission of peroxodisulphate responsible for chain initiation

$$S_2O_8^{2-} \xrightarrow{k_1} 2SO_4^- \tag{1}$$

and the following propagation steps:

$$SO_4^- + H_2O \xrightarrow{k_2} HSO_4^- + OH \tag{2}$$

$$OH + H_2O_2 \xrightarrow{k_3} H_2O + HO_2 \tag{3}$$

$$HO_2 + S_2O_8^{2-} \xrightarrow{k_4} O_2 + HSO_4^- + SO_4^- \tag{4}$$

$$HO_2 + H_2O_2 \xrightarrow{k_5} O_2 + H_2O + OH \tag{5}$$

Step (5) is negligible, as the overall stoichiometry under the conditions of the investigation corresponds to

$$S_2O_8^{2-} + H_2O_2 \rightarrow 2HSO_4^- + O_2$$

Of the six possible bimolecular termination steps only the following are of importance:

$$HO_2 + OH \xrightarrow{k_6} O_2 + H_2O \tag{6}$$

$$SO_4^- + OH \xrightarrow{k_7} HSO_5^- \tag{7}$$

$$HO_2 + SO_4^- \xrightarrow{k_8} O_2 + HSO_4^- \tag{8}$$

$$SO_4^- + SO_4^- \xrightarrow{k_9} S_2O_8^{2-} \tag{9}$$

The rate law can then be derived by employing the steady-state approximation, and by assuming that the rates of the chain initiation and termination steps are negligible compared with the rates of the propagation steps (an assumption justified by the length of the chain).

The expression arrived at, in agreement with that found experimentally, is

$$\frac{-d[S_2O_8^{2-}]}{dt} = \frac{[S_2O_8^{2-}]}{\left[\frac{k_6}{k_1k_3k_4[H_2O_2]}+\frac{k_7[S_2O_8^{2-}]}{k_1k_2k_3[H_2O_2]}+\frac{k_8}{k_1k_2k_4}+\frac{k_9[S_2O_8^{2-}]}{2k_1k_2^2}\right]^{1/2}}$$

Comparison of the empirical and derived rate laws yields a number of relationships between the values of the individual rate constants, e.g., $k_6k_9/k_7k_8 = 9{\cdot}2$, a result which shows that the termination steps are not of equal efficiency.

The results above were obtained at high acidities where the chain-carrying reactions (4) and (5) are pH-independent and where the ionization of the peroxo radical (HO_2) may be neglected. At lower acidities ionization of HO_2 occurs and the chain-carrying species is O_2^- with reactions (4) and (5) replaced by (10) and (11):[16]

$$O_2^- + S_2O_8^{2-} \rightarrow O_2 + SO_4^{2-} + SO_4^- \qquad (10)$$

$$O_2^- + H_2O_2 \rightarrow O_2 + OH^- + OH \qquad (11)$$

together with

$$HO_2 + HO_2^- \rightarrow O_2 + OH^- + OH$$

and, at a high pH,

$$O_2^- + HO_2^- \rightarrow O_2 + OH^- + O^-$$

The catalysed decomposition of hydrogen peroxide

Of all the solution processes involving free radicals the reaction between hydrogen peroxide and iron(II) has received the most attention. The earliest contribution to a knowledge of the mechanism was made by Haber and Weiss,[17] who postulated a chain process involving OH and HO_2 radicals as intermediates:

Initiation:

$$Fe(II) + H_2O_2 \xrightarrow{k_1} Fe(III) + OH^- + OH \qquad (1)$$

Propagation:

$$OH + H_2O_2 \rightarrow HO_2 + H_2O \qquad (2)$$

$$HO_2 + H_2O_2 \rightarrow O_2 + H_2O + OH \qquad (3)$$

Termination:

$$Fe(II) + OH \rightarrow Fe(III) + OH^- \quad (4)$$

The variable stoichiometry of the reaction, depending on the order of mixing of the reactants, is readily explained on such a scheme: a local excess of peroxide would favour propagation of the chain (by (2) and (3)) whereas an excess of iron(II) favours chain termination (by (4)).

Since the rate of decomposition is proportional to iron(III) concentration, showing also an inverse dependence on acidity, the basic scheme is extended by the inclusion of

$$H_2O_2 \rightleftharpoons H^+ + HO_2^- \qquad pK_A = 12 \quad (5)$$

$$Fe(III) + HO_2^- \rightarrow Fe(II) + HO_2 \quad (6)$$

followed by

$$Fe(III) + HO_2 \rightarrow Fe(II) + H^+ + O_2 \quad (7)$$

Reactions (6) and (7) have received support from oxygen-18 tracer experiments which have shown that both oxygen atoms in the oxygen product derive from the same peroxide molecule.

Modifications of the original Haber–Weiss mechanism have been chiefly concerned with the reactive species of the peroxide molecule. The peroxo radical is considered to dissociate by

$$HO_2 \rightleftharpoons H^+ + O_2^- \qquad pK_A \sim 2 \quad (8)$$

and since reaction (3) is unlikely, on the grounds of involving an awkward rearrangement and an improbable transition state, reaction (9) is invoked in its place

$$O_2^- + H_2O_2 \rightarrow O_2 + OH^- + OH \quad (9$$

This step requires only the breakage of the HO—OH bond with electron transfer from O_2^- to OH (cf. reaction (1)). To complete the possibilities, further suggestions have been the electron-transfer steps

$$Fe(II) + HO_2 \rightarrow Fe(III) + HO_2^- \quad (10)$$

and

$$Fe(III) + O_2^- \rightarrow Fe(II) + O_2 \quad (11)$$

Reaction (11) explains the suppression by fluoride ions of the rapid burst of oxygen on mixing peroxide and iron(II) since fluoride effectively complexes all iron(III) formed. A detailed kinetic investigation[18] has allowed k_1 to be defined as $k_1 = 4{\cdot}45 \times 10^8 \exp(-9400/RT)\ M^{-1}\ s^{-1}$.

Evidence for the existence of OH and HO_2 radicals

Apart from the considerable kinetic evidence discussed above for the hydroxyl and peroxo radicals, recent electron spin resonance studies have clearly established their existence. When acidic solutions of titanium(III) and hydrogen peroxide react immediately before passing through the cavity of an ESR spectrometer, two separate singlets are

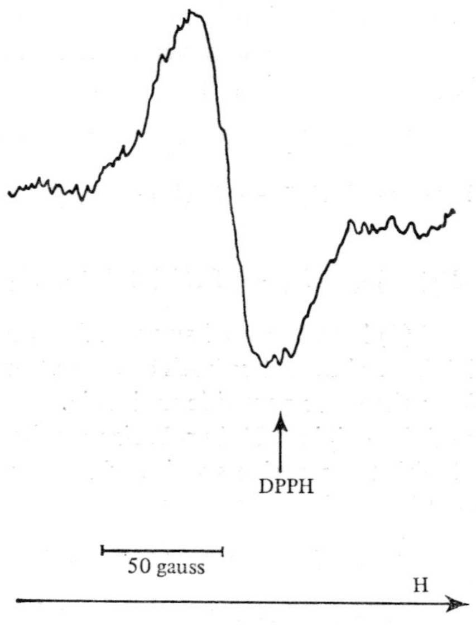

Fig. 5.2. The ESR spectrum of the HO_2 radical produced by interaction of 1×10^{-3} M Ce(IV) with 0·1 M H_2O_2 in 0·8 N sulphuric acid solution (vertical arrow indicates position of DPPH signal). From E. Saito and B. H. J. Bielski, *J. Amer. Chem. Soc.*, 1961, **83**, 4467.

observed in the spectrum which are thought to be due to complexes of OH and HO_2 radicals with Ti(IV).[19] In the presence of methanol the singlets are replaced by a triplet which is attributed to the hydroxymethylene radical originating from the abstraction of hydrogen from the C—H bond of the alcohol by the OH radical.

The ESR spectrum of the peroxo radical has been obtained[20] by reacting Ce(IV) with an excess of hydrogen peroxide in a similar flow system

$$Ce(IV) + H_2O_2 \rightarrow Ce(III) + H^+ + HO_2$$

$$Ce(IV) + HO_2 \rightarrow Ce(III) + H^+ + O_2$$

The *g* value for the observed species (which may be $H_2O_2^+$ [21]) is 2·016 with a line width of 27 gauss (Fig. 5.2). In the presence of excess Ce(III) the resulting signal shows a significant decrease in strength as well as a change in line width. The former effect is believed to result from the formation of a diamagnetic complex of Ce(III) and HO_2, or a back-reaction.

References

1. P. D. Bartlett and J. D. Cotman, *J. Amer. Chem. Soc.*, 1949, **71**, 1419.
2. I. M. Kolthoff and I. K. Miller, *J. Amer. Chem. Soc.*, 1951, **73**, 3055.
3. S. Fronaeus and C. O. Östman, *Acta chem. Scand.*, 1955, **9**, 902.
4. I. M. Kolthoff, P. R. O'Connor, and J. L. Hansen, *J. Polymer Sci.*, 1955, **15**, 459.
5. J. W. L. Fordham and H. L. Williams, *J. Amer. Chem. Soc.*, 1951, **73**, 4855; I. M. Kolthoff, A. I. Medalia, and H. P. Raaen, *J. Amer. Chem. Soc.*, 1951, **73**, 1733.
6. J. W. L. Fordham and H. L. Williams, *J. Amer. Chem. Soc.*, 1955, **77**, 3715.
7. R. Woods, I. M. Kolthoff, and E. J. Meehan, *J. Amer. Chem. Soc.*, 1963, **85**, 2385; *Inorg. Chem.*, 1965, **4**, 697.
8. R. Woods, I. M. Kolthoff, and E. J. Meehan, *J. Amer. Chem. Soc.*, 1963, **85**, 3334.
9. E. Ben-Zvi and T. L. Allen, *J. Amer. Chem. Soc.*, 1961, **83**, 4352.
10. A. J. Kalb and T. L. Allen, *J. Amer. Chem. Soc.*, 1964, **86**, 5107.
11. C. H. Sorum and J. O. Edwards, *J. Amer. Chem. Soc.*, 1952, **74**, 1204.
12. I. M. Kolthoff, E. J. Meehan, and E. M. Carr, *J. Amer. Chem. Soc.*, 1953, **75**, 1439.
13. J. O. Edwards, A. R. Gallopo, and J. E. McIsaac, *J. Amer. Chem. Soc.*, 1966, **88**, 3891.
14. C. E. H. Bawn and D. Margerison, *Trans. Faraday Soc.*, 1955, **51**, 925.
15. M. S. Tsao and W. K. Wilmarth, *Disc. Faraday Soc.*, 1960, **29**, 137.
16. M. S. Tsao and W. K. Wilmarth, in *Advances in Chemistry Series* (ed. R. F. Gould), No. 36, p. 113, American Chemical Society, 1962.
17. F. Haber and J. Weiss, *Proc. Roy. Soc.* (*London*), 1934, **A147**, 332.
18. W. G. Barb, J. H. Baxendale, P. George, and K. R. Hargrave, *Trans. Faraday Soc.*, 1951, **47**, 462; 1951, **47**, 591.
19. Y. S. Chiang, J. Craddock, D. Mickewich, and J. Turkevich, *J. Phys. Chem.*, 1966, **70**, 3509.
20. E. Saito and B. H. J. Bielski, *J. Amer. Chem. Soc.*, 1961, **83**, 4467.
21. B. H. J. Bielski, private communication.

Bibliography

D. A. House, Kinetics and Mechanism of Oxidations by Peroxydisulphate, in *Chemical Reviews*, 1962, **62**, 185.

Free Radicals in Inorganic Chemistry, *Advances in Chemistry Series* (ed. R. F. Gould), No. 36, American Chemical Society, 1962.

6. Protolytic reactions

A knowledge of the rates and mechanisms of protolytic and hydrolytic reactions has come about through the development of chemical relaxation methods. These very rapid reactions had previously proved inaccessible to investigation even by flow techniques. In fact, the successful measurement of the rates of the elementary steps of such processes (notably by Eigen and De Maeyer) has meant a radical reassessment of the term 'instantaneous' as applied to chemical reactions. This chapter describes briefly the reaction between H^+ and OH^-, a reaction which is basic to the understanding of all acid–base equilibria. Using this as a prototype, other processes involving H^+ and OH^- are then considered.

The reaction between H^+ and OH^- ions

The recombination of H^+ and OH^-

$$H^+ + OH^- \underset{k_D}{\overset{k_R}{\rightleftharpoons}} H_2O$$

is the fastest reaction possible in aqueous solution. The rate constant, k_R, has been measured by the dissociation field effect method[1] as $1{\cdot}4 \times 10^{11}$ M^{-1} s^{-1} at 25°, a value corresponding to the theoretical limit for the rate of a chemical reaction. This limit is set by the process of physical diffusion. The reverse reaction, the dissociation of water, has a rate constant, k_D, of $2{\cdot}5 \times 10^{-5}$ s^{-1} at 25°. The ionic product of water, K_W, is related to k_D and k_R by $K_W = k_D[H_2O]/k_R$ where $[H_2O] = 55$ M.

If two ions react spontaneously on collision then the rate of reaction is decided by their rate of diffusion towards one another. For such a situation the reaction is said to be *diffusion-controlled.*

From classical electrostatic arguments (along with the assumption that Stokes' law, for the viscous drag experienced by a macroscopic sphere moving through a liquid medium, applies also to ions) the maximum rate constant (k_R) for the diffusion-controlled recombination of two ions is given by the Debye equation

$$k_R \sim \frac{4\pi N_0 Z_1 Z_2 e^2(D_1 + D_2)}{10^3 \epsilon kT[\exp(Z_1 Z_2 e^2/\epsilon kT\sigma) - 1]} \qquad (6.1)$$

Here, N_0, the Avogadro number, is $6{\cdot}02 \times 10^{23}$ molecules mole^{-1}; Z_1, Z_2 are the integral charges on the ions; $e = 4{\cdot}80 \times 10^{-10}$ esu; D_1, D_2 are the diffusion coefficients of the ions in cm^2 s^{-1}; ϵ is the dielectric constant of the medium; $k = 1{\cdot}38 \times 10^{-16}$ erg deg^{-1}; T is the absolute temperature; and σ is the effective reaction distance (i.e., the distance of closest approach of the two ions) in centimetres. It should be noted that the relationship is strictly applicable only in the case of very dilute solutions. For most ions in dilute aqueous solution σ is about $7{\cdot}5 \times 10^{-8}$ cm (7·5 Å); ϵ is taken to be the bulk dielectric constant of the medium ($\epsilon = 78{\cdot}5$ at 298°K). The values for the diffusion coefficients are obtained from ionic mobilities (in turn derivable from equivalent ionic conductances). Using $H^+ + OH^-$ as an example, $D_{H^+} = 9{\cdot}28 \times 10^{-5}$ cm^2 s^{-1} and $D_{OH^-} = 5{\cdot}08 \times 10^{-5}$ cm^2 s^{-1} at 298°K. On substitution of these values in eq. (6.1) the maximum recombination rate is given by

$$k_R \sim 8{\cdot}80 \times 10^{14}(D_1 + D_2) \sim 1{\cdot}3 \times 10^{11}\ \text{M}^{-1}\ \text{s}^{-1} \text{ at } 298°\text{K}$$

This calculated result compares very favourably with the experimental value of $1{\cdot}4 \times 10^{11}$ M^{-1} s^{-1}. Obviously agreement between theory and experiment depends largely on the value assigned to the reaction distance, σ. However, in other systems the use of 7·5 Å for σ gives rise to values of k_R compatible with the experimentally determined rates. For example, in the reaction

$$H^+ + F^- \underset{k_D}{\overset{k_R}{\rightleftharpoons}} HF$$

k_R is $\leqslant 9{\cdot}4 \times 10^{10}$ M^{-1} s^{-1} by calculation and $\sim 1{\cdot}0 \times 10^{11}$ M^{-1} s^{-1} by experiment. Furthermore, from the exponential nature of the term containing σ in eq. (6.1), k_R is not very sensitive to changes in the value of σ (if σ were as low as 3 Å then k_R would only be reduced to $\leqslant 9 \times 10^{10}$ M^{-1} s^{-1}).

The value of 7·5 Å for the distance of closest approach of H^+ and OH^- shows clearly that the reactive entities are more complex than the simple H_3O^+ and OH^- ions. In fact this result (together with other evidence) is indicative of the species $H_9O_4^+$ and $H_7O_4^-$. These ions have radii of about 4 Å and they can be thought of as the hydronium ion (H_3O^+) and the hydroxide ion in stable association (by hydrogen bridging) with three water molecules. The mechanism describing the interaction of $H_9O_4^+$ and $H_7O_4^-$ can be considered in terms of three stages:

(*a*) the diffusion together of the two hydrated species under the influence of their mutual electrostatic attraction,
(*b*) the formation of a hydrogen bridge between the two species when the ions have diffused to within 6–8 Å of one another, and
(*c*) the rapid transfer of a proton, by rearrangement of the hydrogen bond linkage, to give a neutral ice-like structure which is then broken down by rupture of the hydrogen bonds.

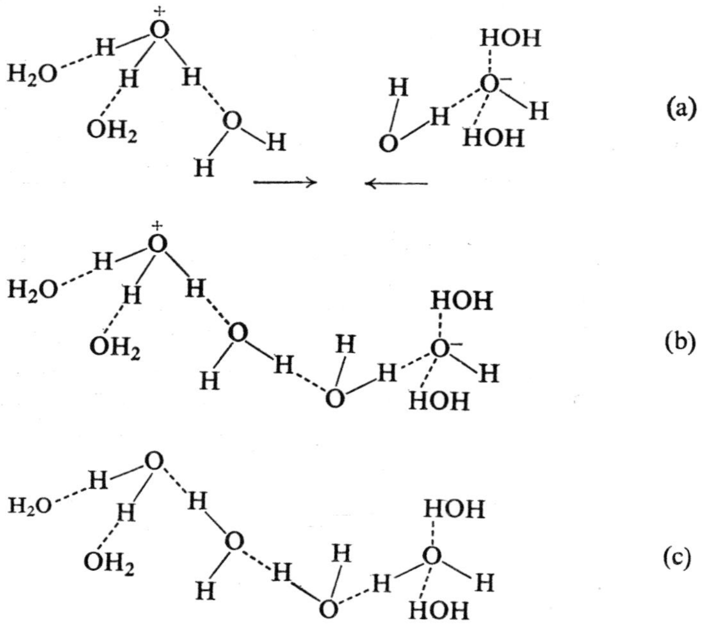

(a)

(b)

(c)

The reactions of the hydrated electron are much slower than those of the hydrated proton. This rather unexpected result is viewed as arising from basic differences in the mechanisms of electron and proton transfer. The exceptionally rapid rates of protolytic reactions are due not only to the high mobility of the proton but also to the existence of hydrogen bridges which act as mediators for the transfer of the proton to an electron localized on the base. The hydrated electron is, by nature, delocalized and therefore less able to recombine. For example, the rate constant for the reaction between H^+ and e^-_{aq} is $2{\cdot}3 \times 10^{10}\ M^{-1}\ s^{-1}$, an order of magnitude less than that for $H^+ + OH^-$. The disappearance of e^-_{aq} by reaction with water

$$e^-_{aq} + H_2O \rightarrow H + OH^-$$

is very much slower than any protolytic reaction.

Other reactions involving H^+ and OH^- ions

As the rate-controlling stage in the $H^+ + OH^-$ reaction is the diffusion of ions and not the actual transfer of a proton, a prediction can be made about the rates of other acid–base reactions. Provided that the reacting base B is a better proton acceptor than water, the rates of all normal protolytic reactions are expected to be diffusion-controlled also. Of course the rates should not be as high as the rate of combination of H^+ and OH^- since these ions have exceptionally high mobilities. It turns out that the rates of recombination are indeed similar (10^{10} to $10^{11}\ M^{-1}\ s^{-1}$). Some typical results are given in Table 6.1. Further-

Table 6.1

Recombination and dissociation rate constants (at 25° and $\mu = 0$) in acid equilibria *

Reaction	k_R, $M^{-1}\ s^{-1}$	k_D, s^{-1}
$H^+ + F^- \rightleftharpoons HF$	$1{\cdot}0 \times 10^{11}$	7×10^7
$H^+ + SO_4^{2-} \rightleftharpoons HSO_4^-$	$\sim 1{\cdot}0 \times 10^{11}$	$\sim 1{\cdot}0 \times 10^9$
$H^+ + HS^- \rightleftharpoons H_2S$	$7{\cdot}5 \times 10^{10}$	$4{\cdot}3 \times 10^3$
$H^+ + HCOO^- \rightleftharpoons HCOOH$	$\sim 5 \times 10^{10}$	$\sim 8{\cdot}6 \times 10^6$
$H^+ + HCO_3^- \rightleftharpoons H_2CO_3$	$4{\cdot}7 \times 10^{10}$	$\sim 8 \times 10^6$
$H^+ + CH_3COO^- \rightleftharpoons CH_3COOH$	$4{\cdot}5 \times 10^{10}$	$7{\cdot}8 \times 10^5$

* From M. Eigen *et al.* (see bibliography).

more, as might be anticipated, the rate is influenced by the charge and shape of the ions. Oppositely-charged anions of high spherical symmetry react the fastest. Examples are F^- and HS^-. Since the point of attachment of a proton is at the site of a lone-pair of electrons it is reasonable that the highly spherosymmetrical F^- ion with four lone electron-pairs reacts faster than the HS^- ion with only three lone pairs. For reactions with H^+, increase of positive charge on the reactant leads to a slight reduction in rate, e.g., the rate decreases in the order $Cu(OH)^+ > [Co(NH_3)_5OH]^{2+} > [Pt(en)_2(en\text{-}H)]^{3+}$. For reactions with OH^-, increase of negative charge on the reactant leads also to a slight reduction in rate, e.g., in the series of mono-protonated phosphates $HPO_4^{2-} > HP_2O_7^{3-} > HP_3O_{10}^{4-}$. Inspection of the data in Tables 6.1 and 6.2 reveals that the reactions of H^+ are much faster,

Table 6.2

Recombination and dissociation rate constants (at 20°) in base equilibria *

Reaction	k_R, $M^{-1}\ s^{-1}$	k_D, s^{-1}
$OH^- + NH_4^+ \rightleftharpoons NH_3 + H_2O$	$3{\cdot}4 \times 10^{10}$	6×10^5
$OH^- + HCO_3^- \rightleftharpoons CO_3^{2-} + H_2O$	$\sim 6 \times 10^9$	. . .
$OH^- + HCrO_4^- \rightleftharpoons CrO_4^{2-} + H_2O$	$\sim 6 \times 10^9$	. . .
$OH^- + HPO_4^{2-} \rightleftharpoons PO_4^{3-} + H_2O$	$\sim 2 \times 10^9$	$\sim 2 \times 10^7$

* From M. Eigen *et al.* (see bibliography).

in general, than those of OH^-. This is, to a large extent, a consequence of the much higher mobility of H^+ relative to OH^-. A simple relationship exists between the strength of the acid and the rate of dissociation, k_D in the equation

$$BH^+ \underset{k_R}{\overset{k_D}{\rightleftharpoons}} B + H^+$$

This stems from the observation that, for a series of acids, the values of k_R, being diffusion-controlled, are of the same order of magnitude and thus the acid dissociation constants, defined as $K_A = k_D/k_R$, are proportional to k_D. In effect this means that strong acids will dissociate rapidly and that weak acids will lose their protons more slowly and to a lesser extent. Similarly in equilibria of the type

$$B + H_2O \underset{k_R}{\overset{k_D}{\rightleftharpoons}} BH^+ + OH^-$$

the k_R values are roughly comparable and a strong base loses hydroxide ions more rapidly than a weak base.

The hydration of carbon dioxide

The hydration–dehydration equilibrium of carbon dioxide has in the past been written as

$$CO_2 + H_2O \rightleftharpoons H_2CO_3 \rightleftharpoons H^+ + HCO_3^-$$

Investigations by both the temperature-jump and pressure-jump methods have indicated that the bicarbonate ion, rather than being derived solely from the deprotonation of carbonic acid, is more likely to be an intermediate

$$CO_2 + H_2O \underset{}{\overset{(A)}{\rightleftharpoons}} H^+ + HCO_3^- \underset{}{\overset{(B)}{\rightleftharpoons}} H_2CO_3$$

On this mechanism the HCO_3^- ion can be protonated in two ways. Proton attack may occur either at the hydroxyl oxygen (*A*) or at the negatively-charged oxygen (*B*):

$$H{-}\ddot{O}{-}C({=}O){-}\ddot{O}{:}^- + H^+ \xrightarrow[B]{} H{-}O{-}C({=}O){-}O{-}H$$

$$+ H^+ \xrightarrow{A} H_2O + O{=}C{=}O$$

Step *A* is considered to be slower than step *B* since the linear molecule CO_2 is formed from the bent HCO_3^- ion, a process likely to require both molecular and electronic rearrangements. In the case of the hydration of SO_2 the structural changes involved are less drastic and the rate is much higher.

Hydrolysis of the halogens

The temperature-jump method has been used to study the hydrolysis of chlorine, bromine, and iodine.[2] The overall reaction, in which a hypohalous acid is formed, corresponds to

$$X_2 + H_2O \underset{k_2}{\overset{k_1}{\rightleftharpoons}} XOH + H^+ + X^-$$

Values for the overall rate constants, k_1 and k_2, are given in Table 6.3. The actual mechanism itself is complex and has been resolved into a series of interrelated stages

$$\begin{array}{ccc} X_2 + H_2O & \underset{k_b}{\overset{k_a}{\rightleftharpoons}} & X_2OH^- + H^+ \rightleftharpoons XOH + H^+ + X^- \\ \updownarrow & & \\ X_2 + OH^- + H^+ & \rightleftharpoons & X_2OH^- + H^+ \end{array}$$

The intermediate X_2OH^- has been postulated also in the exchange reactions of hypochlorite (p. 155). At first sight the overall process is

Table 6.3

(Overall) rate constants (at 20° and $\mu = 0{\cdot}1$ M) for hydrolysis of the halogens:

$$X_2 + H_2O \underset{k_2}{\overset{k_1}{\rightleftharpoons}} XOH + H^+ + X^-$$

X_2	k_1, s^{-1}	k_2, $M^{-2}\ s^{-1}$
Cl_2	11·0	$1{\cdot}8 \times 10^4$
Br_2	111	$1{\cdot}6 \times 10^{10}$
I_2	3·0	$4{\cdot}4 \times 10^{12}$

* From ref. (2).

hardly recognizable as a protolytic reaction. However, one of the steps (k_b) is seen to be a protolysis (the reverse step, k_a, is a hydrolysis). Characteristically, k_b displays a diffusion-controlled rate common to all three halogens. Values have been obtained for the rate constants of all the individual steps.

References

1. M. Eigen and L. De Maeyer, *Z. Elektrochem.*, 1955, **59**, 986.
2. M. Eigen and K. Kustin, *J. Amer. Chem. Soc.*, 1962, **84**, 1355.

Bibliography

M. Eigen, Proton Transfer, Acid-Base Catalysis etc., Part I: Elementary Processes, *Angewandte Chemie* (*International Edition*), 1964, **3**, 1.

M. Eigen, W. Kruse, G. Maass, and L. De Maeyer, Rate Constants of Protolytic Reactions in Aqueous Solution, in *Progress in Reaction Kinetics* (ed. G. Porter), Vol. 2, p. 285, Pergamon, 1964.

The Kinetics of Proton Transfer Processes, *Discussions of the Faraday Society*, No. 39, 1965.

Subject index

Printed in Great Britain by Spottiswoode, Ballantyne & Co. Ltd.